MUSIC AND SOUND

HERMANN VON HELMHOLTZ
(1821–1894)

MUSIC AND SOUND

By Ll. S. LLOYD, C.B., M.A. (Cantab.)

Llewelyn Southworth

*Principal Assistant Secretary to
the Department of Scientific and Industrial
Research*

With a FOREWORD *by*

Sir WILLIAM BRAGG, O.M., K.B.E., D.Sc.

*President of the Royal Society
Director of the Royal Institution of
Great Britain*

GREENWOOD PRESS, PUBLISHERS
WESTPORT, CONNECTICUT

ML
3805
L5.8 M9
1970a

Originally published in 1937
by Oxford University Press, London

First Greenwood Reprinting 1970

Library of Congress Catalogue Card Number 70-109770
SBN 8371-4260-1

Printed in the United States of America

FOREWORD

LECTURES on acoustics are included in every course that leads to a diploma or a degree in Music, and the lecturer treats the subject as a branch of Physics. Many of us who have been called upon to undertake this duty have felt that we lacked some of the requisites for success. We were accustomed to classes of students whose physical studies were more or less advanced; the musicians were a class apart. We knew very little, that was my case at least, of the intricacies of harmony and counterpoint, and we were unable to make use of illustrations which our hearers would have welcomed.

Mr. Lloyd is better equipped. To his knowledge of acoustics as a branch of physics he adds an understanding of the special difficulties and requirements of students of music, being himself a musician.

It is somewhat of a tragedy that, although he can count on the artist as an ally to an extent which depends on the instrument, the composer who tries to write in a perfect scale fights with the laws of arithmetic in a battle which he can never win. In the adopted compromise of the temperaments accuracy is sacrificed for the sake of freedom. This difficulty and the nature of the compromise and all that it means to music are best handled by a man who is himself familiar with its effects and can answer the questions which the student of music is bound to ask. In his treatment of this very important subject, and in his consideration of all the other cases in which the laws of physical acoustics are of importance in the rendering of music, Mr. Lloyd has begun at the musician's end. He has not based his argument on a development of the principles of dynamics. He has preferred to meet his students on their own familiar ground, and to lead them thence to points from which they can see how the laws of acoustics march with their own territory.

Mr. Lloyd's book is sure to excite the interest of those for whom it has been written, and it is to be expected that his novel way of putting his matter before them will help them very greatly.

W. H. BRAGG

CONTENTS

CONTENTS

INTRODUCTION

THIS book is intended in the first instance for the use of students of music; but it may have interest for others who wish to know how far the science of acoustics can explain the difference of quality in the notes of different musical instruments or why the art of music can use an imperfect scale. The scientific text-book on sound approaches the relation between acoustics and music from the scientific angle. The excuse for the existence of this book is that it approaches the subject from the point of view of the history of musical composition. That is to set the science in its proper place in relation to the art of music, and to assist the student to make, as far as possible, his knowledge of counterpoint, harmony, the development of musical form, the history of music, and acoustics into a consistent whole.

Some musicians are inclined to doubt whether an elementary knowledge of acoustics is of any real value to students of music, and this doubt is clearly justified if the student's knowledge is vague or if he does not understand where it is based on scientific assumption. Moreover, in his article on 'Harmony' in the fourteenth edition of the *Encyclopædia Britannica*, Sir Donald Tovey writes: 'The art of music had not attained to the simplest scheme for dealing with discords before it traversed the acoustic criterion in every direction.' Music is not a science, but this does not mean that some knowledge of acoustics is not useful to the musician. The contrary is true, provided that, so far as it goes, his knowledge is exact. If the development of the scale, with its relation to acoustical phenomena, be correctly apprehended as part of the history of musical composition, the study of some of those aspects of acoustics which have a practical interest for musicians will help to throw additional light on one aspect of musical history.

Nor is this all. In his book, *Musical Composition*, Stanford lays it down at the outset that it is an absolute necessity for the composer to study the pure scale and write in it, and, as a first step, to learn to think in it. To do this he must study music of the polyphonic period when composers of Church music wrote for unaccompanied voices. This must, of course, be done away from the pianoforte; and for any one whose early acquaintance with music was obtained through that instrument the process of thinking in the pure scale is certainly assisted by a clear intellectual perception of the nature of its intervals.

The study of acoustics must always have a real interest for the musician. As the subject unfolds itself to him he cannot but be fascinated to discover what a wonderful organ he possesses in his ear. As will be shown in this book, in listening to a musical note it performs the functions of an expert mathematician and a lightning calculator with an ease which completely conceals its achievement, while it is able to distinguish two narrowly separated vibrations with a certainty to which the eye can attain, in the case of light, only if it be assisted by a spectroscope. It is not surprising, therefore, that as scientific knowledge increased, claims should have been made for its importance for the art of music which the musician instinctively rejects.

From the time of the Greeks the curious-minded have sought to find in nature an explanation of music. The history of musical theory is full of misguided conjecture about acoustics and music which neglected to satisfy the condition that alone permits rational conjecture. This condition is that conjecture must be tested by evidence obtained both from rigorous experimental investigation and from the history of musical composition.

It was Helmholtz who first established a correct basis for the study of the relations between different notes. What matters to the musician is the sensation transmitted from his ear to his brain and nothing else. What happens outside the ear is of no importance to him except in the way it affects his sensations through the ear. As a distinguished physiologist as well as physicist, Helmholtz was exceptionally fitted to solve the problems to which previous investigators had failed to find correct answers. The full title of his classical treatise is, in its English translation, 'On the Sensations of Tone as a physiological basis for the theory of music'. As we shall see in due course, the touchstone of much of his experimental method was the examination of beats; and the unpleasant quality in the sensation of beating is a physiological effect in the ear itself. By his investigations Helmholtz established an exact criterion for acoustical effects as measured by the ear which is as far reaching as it is simple. The test which the European ear applies to a scale system is whether notes which it wishes to hear in tune do, in fact, sound in tune. To this question also Helmholtz's criterion is directly applicable. And it should be added that he himself observed 'that the system of scales, modes, and harmonic tissues does not rest solely upon unalterable natural laws, but is also, at least partly, the result of aesthetical principles, which have already changed, and will still further change, with the progressive development of humanity'.

An illustration of the dangers of unsound conjecture will be found in Chapter V. Certain physical hypotheses due to Rameau and d'Alembert, on which a theory of harmony was constructed, were worked to death, particularly in this country. The conceptions of chords as things in themselves were developed by Dr. Day to a point at which they can only be described as tidy classification run mad. The composer's view of them is stated by Stanford in his book, *Musical Composition*, with convincing finality: 'The growth has unfortunately overrun a great deal of low-lying land, and it is easy enough to note where it flourishes from the results of its miasma.'

It was to correct this error that Stanford advised his pupils to revert to the practice of learning harmony through counterpoint. As Tovey observes: 'The great classical tradition cares little for the study of chords as things in themselves; and the art of harmony perishes under a discipline that separates its details from counterpoint and its larger issues from form.' In short, music is an art, not a science.

If the student approaches the subject with this firmly in mind he will find that there is much in acoustics which, if clearly grasped, will fit into his other studies. This explains the aim of this book. It begins with a study of the scale: it proceeds to the study of dissonance which is significant for the appreciation of the development of the scale; and it goes on to inquire into the nature and properties of sound only when knowledge of that subject is required fully to understand the cause of quality in the musical notes produced by different instruments. So far as it goes it attempts to do this thoroughly. If the attempt has been successful, this book may be a useful complement to scientific text-books on sound which treat the subject more fully.

In approaching the study of acoustics the student of music has a great advantage over students of science who may have had no musical training. He does not need to be told that a musical note is produced by regular vibrations such as those of vibrating strings or tuning-forks, that the more rapid the vibration the higher the pitch of the note, that the greater the vibration the louder is the note. He is probably acquainted with the positions of the harmonics of a note; he has a practical perception of the sensation of beating; and he recognizes the nature of the difference of quality in the same note as produced by different instruments or, say, the different stops of an organ. He has learnt from counterpoint which intervals and combinations of intervals are treated as concords by the art of music. I believe that the method of approach adopted in this book is the logical one for students of music who may know little science.

It is impossible to give a complete account of the principles of acoustics without some recourse to mathematics. The science of dynamics, which deals with forces, masses, velocities, and accelerations, is required to determine the laws of vibrating bodies. In this book the conclusions of dynamics are accepted. In the chapters which deal with the physical aspect of the subject the plan adopted has been to give elementary indications of the methods employed by the mathematician. These have been placed in appendices, so that non-mathematical readers can ignore them if they wish or merely note their conclusions. But by using graphical methods I have attempted to induce the non-mathematical reader to find out for himself *how* sounding bodies vibrate and *how* air vibrates when disturbed by sound, leaving it to the mathematician to show *why* they vibrate as they do—a matter of no immediate importance to the musician. And while it is impossible, without using logarithms, to represent any scales graphically or to calculate the intervals employed in tuning keyboard instruments, I have attempted to use them in an elementary way which can be mastered by any student with a schoolboy's knowledge of algebra; though readers will find that, as explained in Chapter I, they will be able to understand this book without becoming acquainted with logarithms.

In conclusion, I desire to acknowledge my great indebtedness to Sir William Bragg, for reading my text critically, for making numerous valuable suggestions, and for providing a Foreword. I have to thank two colleagues, Dr. Kaye for suggesting the treatment of reflection in Chapter VII and for furnishing the note on overblowing of organ pipes in Chapter IX, Dr. Davis for calculating for me the magnitude of the vibration reaching the ear which is given on p. 72 and the reflection factors given on p. 89, and both of them for much of the information contained in Appendix XI. My special thanks are also due to Dr. Kaye for the hitherto unpublished photograph of Helmholtz which is reproduced as the frontispiece of this volume. Dr. Kaye and other colleagues, Dr. Aston and Mr. R. O'F. Oakley, have undertaken the task of reading the proofs. Finally, I desire to record my obligation to Sir Percy Buck for the kind interest he has taken in my text and for allowing me to consult him about my description and treatment of the modern scale system.

LL. S. LLOYD

HELMHOLTZ'S PITCH NOTATION

In Helmholtz's pitch notation the notes proceeding upwards from C are called D, E, F, G, A, B, c, d, e, f, g, a, b, c′, d′, e′, f′, g′, a′, b′, c″, and so on. Similarly the notes between C and C, all carry the suffix ,
and so on.

'*Helmholtz, by a series of daring strides, has effected a passage for himself over that untrodden wild between acoustics and music—that Serbonian bog where whole armies of scientific musicians and musical men of science have sunk without filling it up.*'

CLERK MAXWELL, *Rede Lecture, 1878.*

I

PRELIMINARIES, THE PURE SCALE, EASY LOGARITHMS

'IT is advisable to guard at the outset against the familiar misconception that scales are made first and music afterwards. Scales are made in the process of endeavouring to make music, and continue to be altered and modified, generation after generation, even till the art has arrived at a high degree of maturity. The scale of modern harmonic music, which European peoples use, only arrived at its present condition in the last [i.e. the 18th] century, after having been under a gradual process of modification from an accepted nucleus for nearly a thousand years.' No better introduction to the study of the scale could be found than this succinct review of its history which occurs in Chapter II of Parry's *Art of Music*.

All the evidence shows that, in the early stages of a scale developed in the attempt to sing melody, one of two intervals, the fourth as an interval approached downwards, or the fifth, would almost certainly provide its first essential note other than the octave. Other notes would be developed by adding ornamental notes round the essential ones. Whatever the origins of the European Scale, it is certainly the case that Pythagoras (*c.* 570–500 B.C.) made use of the simple divisions of a vibrating string to verify, or as we might say to standardize, the essential intervals of the scale. These gave him the octave, the fifth, and the inversion of the fifth, the fourth. Just as the dominant is a true fifth above the bottom note of the scale, so our subdominant is a true fifth below its octave. There is undoubtedly a natural basis for the octave. As we shall see in Chapter V, this is an interval hedged in and defined by the sharpest dissonances. It gives a unique sense of identity of the two notes. It is possible to confuse for a moment the intervals of two and of three octaves, especially if the notes are sounded by different instruments. It is inconceivable that a similar confusion could occur between intervals made up of two fifths and three fifths respectively, for example in the tunings of a violin.

Intervals.

As was noted in the Introduction, musical notes are produced by regular vibrations, such as those of strings or tuning-forks.

The more rapid the vibration the higher is the pitch of the note. The rate of vibration which determines the pitch of a note is conveniently referred to by the descriptive term 'frequency'.

It is proved by experiment in the laboratory that any pairs of notes, the ratios of whose frequencies are the same, are accepted by the ear of the musician as having the same musical relation to one another. This ratio therefore determines, physically, the musical interval. This is a very important fact, which we must consider before proceeding further. But for this fact we could not have scales as we know them, scales which are the same at different pitches and repeat at the octave.

If we sounded in succession three notes which vibrated 100, 150, and 225 times a second respectively, the musician would at once tell us that the first and second were a fifth apart and that the second and third were a fifth apart. The physicist would say that the ratio of the frequency of the second note to that of the first was 150 : 100 or, as a fraction, $\frac{3}{2}$. He would say that the ratio of the frequency of the third note to that of the second was 225 : 150 or, as a fraction, likewise $\frac{3}{2}$.

From this fact there emerges another which is purely a matter of arithmetic. The ratio of the frequencies of the first and third notes is 225 : 100 or, as a fraction, $\frac{9}{4}$. The musician would say that the interval between the first and third notes is what he calls a ninth, and that it is made up of two fifths. The physicist says that the ratio $\frac{9}{4}$ is the product of the ratios which he found for the two fifths, namely $\frac{3}{2} \times \frac{3}{2}$. If we use letters instead of numbers we can readily see that this is a general truth. The mathematician uses letters instead of numbers as a kind of shorthand. When he writes down x he means: 'Think of any number you like, but having chosen your number do not change it. I shall call it x.'

Consider three notes, and suppose that the frequency of the second is x times the frequency of the first and that the frequency of the third is y times that of the second. It is obvious that the frequency of the third is $x \times y$ times that of the first. Or suppose that the frequencies of the three notes in ascending order are a, b, and c. The ratio of the frequencies of the first two is represented by the fraction $\dfrac{b}{a}$; that of the frequencies of the second two by the fraction $\dfrac{c}{b}$; while that of the frequencies of the third and the first is represented by $\dfrac{c}{a}$, which equals $\dfrac{b}{a} \times \dfrac{c}{b}$.

Hence the rule that to determine, physically, the sum of two musical intervals which have a note in common we multiply the ratios of the frequencies of the two notes which in each case form the interval. Such a ratio we may conveniently refer to as the ratio of the interval. Similarly if we wish to determine, physically, the difference between two intervals which have a note in common we divide the ratios of the intervals. *To add intervals multiply their ratios. To take one interval from another, divide their ratios.*

Notes whose frequencies are represented by simple ratios such as $\dfrac{2}{1}$, $\dfrac{3}{2}$, or $\dfrac{4}{3}$, produce musical intervals which are pleasing. These three fractions are, in fact, the ratios of the intervals of an octave, a fifth, and a fourth respectively. These were the only intervals accepted by the Greeks as consonances. Even the major third they regarded as a dissonance.

The Harmonic Series.

A flexible string can vibrate as a whole, as two half-lengths, three thirds of a length, and so on; and it can execute all these vibrations simultaneously. The vibration as a whole gives the fundamental note of the string. If the string is uniform and very flexible, the sectional vibrations give rise to harmonics or overtones falling into a series which can be represented in staff notation thus:

Fig. 1

The notes in this series have frequencies in the ratios 1, 2, 3, 4, &c., as is proved by laboratory experiment; and the series is called the harmonic series.[1] The B♭, shown as the seventh note of the series, is represented by a black note because this note of the series is actually flatter than the minor seventh above C. In Chapter IV we shall see how we may ascertain its exact position. It is not quite correctly to be represented by any note that can be shown on the stave.

The notes in the harmonic series are familiar to many musicians. They are therefore given here because they are useful to refer to as a check on memory when quoting the ratio of any of the simpler intervals, such as the major third which has the ratio 5 : 4, or the minor third which has the ratio 6 : 5, or the fifth with the ratio 3 : 2, or the major sixth with the ratio 5 : 3.

A word of warning must be given. The student may observe that in the harmonic series with C as its fundamental he can find the notes of the major scale of C except the fourth and the sixth, F and A. Thus the interval between the fundamental and the second degree, C to D, will be found between the notes of the series marked 8 and 9, that for the third degree, C to E, between those marked 4 and 5, and that for the fifth, C to G, between those marked 2 and 3, while if he were to continue the series upwards he would find that the interval of the major seventh, C to B, was given by the eighth and fifteenth notes of the series. To conclude from this that the relationships of the harmonic series, which is an acoustical series, may account directly for the intervals of the scale would, in fact, be to indulge in that unsound conjecture against which a warning was given in the Introduction. In Chapter V the student will find an explanation of how some of these relationships have assisted indirectly to fix the intervals of the pure scale.

The overtones of vibrating strings are not perceived by the unpractised ear, while the upper ones are not readily detected even by the practised ear. How far, and in what way, overtones other than the first two or three may have influenced the un-tutored practitioner in the selection of the intervals of his scale at a time when all music was melodic, are questions to which natural science can make no definite answers.

Pythagoras, determining the intervals of the Greek modes by a series of fifths (as will be explained on p. 32), obtained a third which is faulty for harmonic use. His interval was one whose

---

[1] The ratios of these vibrations were published by Sauveur in 1701 (when J. S. Bach was sixteen years old).

ratio was, in fact, $\frac{81}{64}$. Five centuries later Didymus substituted

an interval which, in fact, had a ratio $\frac{5}{4}$. The theoretical differ-

ence is not likely to have troubled the singers of Greek melody in the practice of their art. The leap of a third or a sixth requires some education before it can be sung correctly. The major sixth, indeed, is so difficult an interval to hit correctly that the Italian composers of the polyphonic period rigidly excluded it from their contrapuntal resource.

Nomenclature: Overtones and Partial Tones.

In the preceding section mention was made of overtones. Before proceeding further it will be convenient to make reference to nomenclature so that exact meanings may be attached to the terms we shall find it necessary to use when we wish to describe the acoustical aspect of a musical note.

In a later chapter, when we come to discuss the quality of musical sounds, we shall find that great importance attaches to the presence, in a note, of the overtones of its fundamental element. At the moment it is sufficient to remember that, as the musician is well aware, the notes of most instruments contain a number of overtones which bear to the fundamental element of the note the relations of notes in the harmonic series. The nomenclature we shall adopt to describe the complex sounds which build up a musical note or *tone* is as follows:—

The simplest constituents, each with its own characteristic frequency, which build up a note are called *pure tones*. A *pure tone* is therefore an element of a note which cannot be further analysed.

When the context prevents ambiguity, as in the next paragraph, we may sometimes drop the adjective. The term tone will then correspond to the German word *ton* which Helmholtz used in a similar sense; and this usage may help a reader who may wish to consult Helmholtz's work for himself.

[1] The ratios discovered by Pythagoras for the octave, the fifth, and the fourth, were those of the lengths of vibrating strings. Only in the last three hundred years have investigations, first those of Mersenne published in his *Harmonie Universelle* in 1636, later those of Newton, Brook Taylor (1715), Daniel Bernoulli (1755), and others, established the laws of vibration of strings by which it is shown that the ratios of the lengths of vibrating strings apply also to the pitch numbers or frequencies of the tones they produce: see *Sensations of Tone*, Helmholtz, Chapter I, and *Theory of Sound*, Rayleigh, §§ 15, 124, and 125. (Palestrina died in 1594; Byrd died in 1623.)

A body which can vibrate freely in various ways produces as its lowest tone the *fundamental* or *prime tone* and as higher tones the *overtones*.

Overtones may be either *harmonic* or *inharmonic*. *Harmonic overtones* have pitches which form the harmonic series with the fundamental. The note of a vibrating string consists of the fundamental and a series of overtones which fall into its harmonic series. The overtones of drums and tuning-forks do not fall into the harmonic series of the fundamental. Such overtones are called *inharmonic*.

It is convenient to group together the fundamental tone of a note and its overtones and call them all *partial tones*. We shall commonly use this term to describe a fundamental and overtones which fall into the harmonic series, though it is quite possible for an instrument to select certain notes only of the harmonic series for its overtones. This happens, in fact, for reasons which we shall find out in Chapter IX, with a stopped organ pipe or an orchestral clarinet. But it will be obvious that, when the full harmonic series is present in the overtones, the first overtone, the octave of the fundamental, is the second partial tone; the second overtone, the twelfth of the fundamental, is the third partial tone; and so on. The numbering of the partial tones is then the same as the numbering which represents the ratios of their frequencies to that of the fundamental, an obvious convenience. But observe that this is not true if notes of the harmonic series are lacking: thus the second partial tone of a stopped organ pipe, which lacks the octave overtone, would be the twelfth of the fundamental, that is, the third note of the harmonic series.

The Pure Scale.

After these digressions, the significance of which will become increasingly evident, we may return to consider the pure scale. This, as was noted in the Introduction on the authority of Stanford, is the scale which singers would use as exhibiting the 'concords' employed by the composer of the polyphonic period.[1]

---

[1] Those readers who may not have made themselves familiar with some sixteenth-century music should know that this music was written contrapuntally; and they may accept Stanford's description of *counterpoint* as concurrent melody. That is why musicians refer to the *polyphonic* period, which reached its golden age in the sixteenth century. Reduced to its simplest terms, by way of contrast, the object of later *harmonic* writing may be described as the accompaniment of a melody by other notes which sounded pleasing or otherwise suitable. These descriptions may help; but as explanations they would be a wholly inadequate substitute for the study of the

We shall find it necessary to give names to the notes of the scale. To refer to them as tonic, supertonic, mediant, &c., will be unduly cumbersome. We will call them C, D, E, &c., clearly understanding that the first note, C, may be of any pitch and not that adopted, with more or less exactness, in the tuning of musical instruments. With this understanding we may also use, in a relative sense, Helmholtz's pitch notation, shown in terms of staff notation on p. 1; for example, to show the octave of C as c whatever the exact pitch, and the note below C as B,.

Calling our bottom note C, regardless of its exact pitch, the first thing is to find c, which has double the frequency of C. The ratio of the interval cC is therefore $\frac{2}{1}$. The next thing is to find G, the fifth above C, which has with C the ratio $\frac{3}{2}$. The next is to find F, which is a fifth below c and has with C the ratio $\frac{2}{1} \div \frac{3}{2}$, or $\frac{4}{3}$. (The convenience of notes of the harmonic series as a check on memory of these intervals now appears.) Thus, using the perfect concords of the octave and the fifth, we have found c, G, and F, notes of the scale which would be found by the method of Pythagoras. As the dominant and subdominant, G and F are to-day the most important notes of the scale of C.

For the remaining notes we may use 'concords' on C, F, and G, each consisting of a major third from the bass, with a ratio $\frac{5}{4}$, and a perfect fifth from the bass with a ratio $\frac{3}{2}$. Beginning with this major triad on C we find another note of the scale, E, with the ratio $\frac{5}{4}$ to C. Taking the major triad on F we find A, which has the ratio $\frac{5}{4}$ with F, and therefore the ratio $\frac{5}{4} \times \frac{4}{3}$, or $\frac{5}{3}$, with C. Taking the major triad on G we find two more notes: B with the ratio $\frac{5}{4}$ to G and d with the ratio $\frac{3}{2}$ to G. Their ratios with C are therefore

technique of, say, Palestrina, Bach, and Beethoven, and of the use they made of musical 'resource'. The available concords of three-part counterpoint will be found on pp. 64 and 65.

$\frac{5}{4} \times \frac{3}{2}$, or $\frac{15}{8}$, and $\frac{3}{2} \times \frac{3}{2}$, or $\frac{9}{4}$, respectively. Consequently D, which is an octave below d, has a ratio with C of $\frac{9}{4} \div 2$, or $\frac{9}{8}$. We have now found a scale of seven notes which we can show diagrammatically thus, with the ratios of their frequencies to that of C. In this diagram and in those that follow the intervals are drawn to scale as explained on pp. 19 and 20.

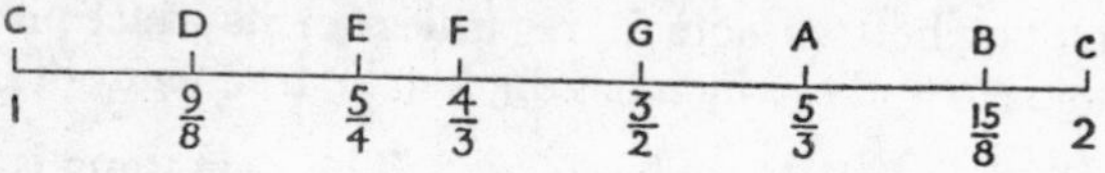

FIG. 2

From these we can calculate the ratios of the intervals between successive notes of the scale and we find there are three with the ratio $\frac{9}{8}$, namely: DC, GF $\left(\frac{3}{2} \div \frac{4}{3}\right)$ and BA $\left(\frac{15}{8} \div \frac{5}{3}\right)$; two with the ratio $\frac{10}{9}$, namely: ED $\left(\frac{5}{4} \div \frac{9}{8}\right)$ and AG $\left(\frac{5}{3} \div \frac{3}{2}\right)$; and two with the ratio $\frac{16}{15}$, namely: FE $\left(\frac{4}{3} \div \frac{5}{4}\right)$ and cB $\left(2 \div \frac{15}{8}\right)$. The three intervals, whose ratios are $\frac{9}{8}$, $\frac{10}{9}$, and $\frac{16}{15}$, are called the major tone, the minor tone, and the (diatonic)[1] semitone respectively. It is unfortunate that the scientist should have chosen the same word 'tone' to describe a musical note or its elements as the musician uses to describe an interval. But in practice no confusion is likely to occur in the mind of the musician.

The major tone is often called the greater tone and the minor tone the lesser tone. This has some advantage, because, as is evident from their ratios, the difference between a major tone and a minor tone is much less than the difference between a major third and a minor third or, what is the same thing, that between a major sixth and a minor sixth. But the consistent use of the Latin terms serves to remind us that the major sixth is the greater sixth. Confusion of the two sixths sometimes occurs because we also use the Latin terms with quite a different meaning, to describe the major and minor keys, when they become what are called terms of art.

[1] *diatonic*, derived from the Greek modes, is a musical term used in contradistinction to *chromatic*.

If now we represent the major tone by $>$, the minor tone by $\Rightarrow$, and the semitone by $-$, we can show the scale diagrammatically thus, when it is evident that the major and minor tones occur alternately except at the octave:

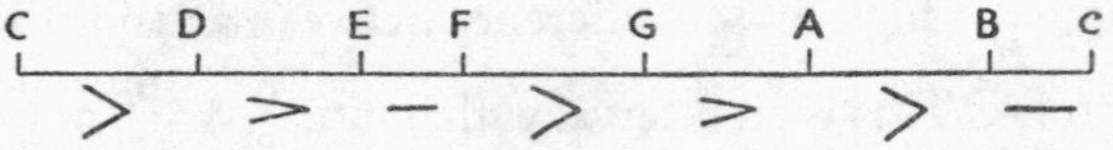

FIG. 3

and we can say that the octave of the pure scale consists of three major tones, two minor tones, and two semitones; the fifth of two major tones, a minor tone, and a semitone; the major third of a major tone and a minor tone; the minor third of a major tone and a semitone; and so on.

It will at once be observed that A is not a major tone above G. If therefore we want to move into the key of G we need a new note, which for the moment we may call A+, to give the relationship we found for the first four notes of the scale of C, namely:

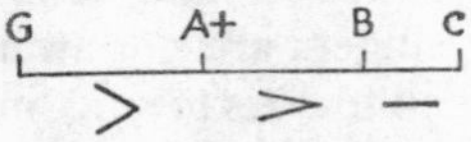

FIG. 4

In addition, of course, we require a new note in place of F, a semitone below G. Owing to the limitations of staff notation we call this new note F♯; but it has nothing to do with F.

The same process continues as we construct the scales for successive fifths D, A, E, &c. Each time the frequency of one note is slightly increased and one note is replaced by a new note, or, as we say, sharpened. Before long few notes of the original scale will remain in use. We may remind ourselves that we are still thinking of a pure scale for music that is sung.

A similar thing happens if we move in the opposite direction to what are called flat keys. A new note altogether is wanted in place of B. For the four notes beginning with F we want a succession—major tone, minor tone, semitone; whereas the notes F, G, A, and B give us a succession—major tone, minor tone, major tone. The new note to replace B must be a semitone above A. It is called B♭, but it has nothing to do with B. Observe that B♭ is a major tone below c, for we have simply reversed the order of the intervals between A and c in the pure scale of C. Proceeding upwards we find that d will not do for the scale of F. The note we require should be a major sixth, ratio $\dfrac{5}{3}$, or two major tones, two minor tones, and a semitone, above F. Actually d has a ratio $\frac{3}{2} \times \frac{9}{8}$, or $\frac{27}{16}$, with F and the interval is made up of three major tones, a minor tone, and a semitone. d is therefore too high by the difference between a major tone and a minor tone. This is an interval with a ratio $\frac{9}{8} \div \frac{10}{9}$, or $\frac{81}{80}$, and it is called a *comma*. More particularly it might be called the comma of Didymus, for it is the amount by which Didymus corrected the major third as established by Pythagoras, as we have seen. We might have obtained it directly by calculation; for $\frac{27}{16} \div \frac{5}{3}$ gives us $\frac{81}{80}$.

Thus whether we modulate up or down the notes of the pure scale by fifths, we find that some notes are either replaced by new notes, which we call sharps or flats, or displaced by intervals of a comma. Certainly singers are not aware of commas as such: what they are concerned to do is to sing 'in tune' with each other, and composers who have trained themselves by the discipline of strict counterpoint write music for them which is grateful to sing in tune.

The effect of modulation in displacing the notes of the scale may be illustrated by comparing two somewhat crude changes of key as follows:

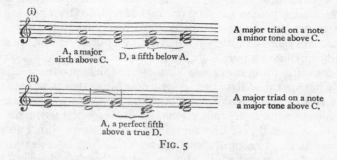

(i)

A, a major sixth above C.   D, a fifth below A.

A major triad on a note a minor tone above C.

(ii)

A, a perfect fifth above a true D.

A major triad on a note a major tone above C.

FIG. 5

In example (i) A is fixed in its true place before D is sounded. When sounded D has then to suffer *mutation*, by a comma, to be a concord with A. The ratio of the major sixth AC is $\frac{5}{3}$. The interval AD must be a concord; and it becomes a perfect fifth with a ratio $\frac{3}{2}$. Consequently the ratio of the interval DC becomes $\frac{5}{3} \div \frac{3}{2}$, or $\frac{10}{9}$, which is the ratio of a minor tone. Observe that C♯ in this example, being a major third above A,, has a frequency ratio with C of $\left(\frac{5}{3} \div 2\right) \times \frac{5}{4}$, or $\frac{25}{24}$, and is the leading note of a D which has suffered mutation. We may check this: $\frac{25}{24} \times \frac{16}{15}$ is equal to $\frac{10}{9}$.

In example (ii) G is sounded in the first chord and continues in the second chord. This requires D, when sounded with it, to be a perfect fourth below G; that is, to be a major tone above C. For $\frac{3}{2} \div \frac{4}{3}$ is $\frac{9}{8}$. As a consequence, A when sounded has to conform to a true D as bass, and to be a perfect fifth above it. This requires that A shall undergo the adjustment required for modulation into a sharp key, for example the key of G, as already seen on page 11 when we called the altered note A+. We may check this by calculation. The ratio of the interval AC will become $\frac{9}{8} \times \frac{3}{2}$, or $\frac{27}{16}$, a fraction which we have already found for an interval of a major sixth plus a comma. A has thus to be altered to a note a comma higher.

The first three chords in the first example also illustrate the fact that in the pure scale of C itself, D has to be a mutable note so as to form a perfect fifth with A when required. This is due to what Parry describes as 'the curiously paradoxical facts of acoustics which make an ideally perfect scale impossible'. The fundamental fact is that successive tunings in fifths can never coincide with successive tunings in octaves which begin with the same note. This is made evident by working out two series of frequencies, each beginning at 100, as follows:

by octaves 100, 200, 400, 800, 1,600, 3,200, &c.
by fifths 100, 150, 225, $337\frac{1}{2}$, $506\frac{1}{4}$, $759\frac{3}{8}$, &c.

The reason is an arithmetical one: the numbers 3 and 2 are prime numbers, a prime number being a number which is not divisible without remainder by any other number except unity. The same result must be produced by successive tunings in thirds; for 5, 3, and 2 are all primes to one another.

We may note, in passing, that in example (ii) the frequency of C♯ has a ratio $\dfrac{135}{128}$ to that of C. Here the octave of C♯ is a perfect third above the note which we called A+ on page 11. The ratio of the interval C♯ C is therefore $\left(\dfrac{27}{16}\times\dfrac{5}{4}\right)\div 2$, or $\dfrac{135}{128}$. This gives a C♯ which is a semitone below a true D, for $\dfrac{9}{8}\div\dfrac{16}{15}$ is $\dfrac{135}{128}$.

When we come to adapt the pure scale for use in the minor, certain notes need alteration. Consider the notes between A, (a minor third below C) and A. Two important notes will be the dominant, a fifth above A, and the subdominant, a fifth below A. In the pure scale of C the interval from A, to E contains two major tones, a minor tone, and a semitone. It is therefore a perfect fifth. But the interval from D to A contains one major tone, two minor tones, and a semitone. It lacks a comma to make it a perfect fifth. Consequently in the minor scale D must be moved down a comma, when it becomes a note a minor tone above C, ratio $\dfrac{10}{9}$. Again, G is a minor tone below A. Melodically it should be a major tone below A and harmonically it should be a true minor seventh above A,. We know that in the major scale of C the interval FG, is a minor seventh, and it is the sum of two perfect fourths: the ratio of a minor seventh is therefore $\dfrac{4}{3}\times\dfrac{4}{3}$, or $\dfrac{16}{9}$. G must be moved down a comma for use in the minor scale. It will then be a semitone plus a minor tone above E, and the ratio of the interval GA, will be $\left(\dfrac{3}{2}\times\dfrac{16}{15}\times\dfrac{10}{9}\right)$, or $\dfrac{16}{9}$, the ratio of a minor seventh.

We can now show the minor scale on A, diagrammatically, with the relative major below it as follows:

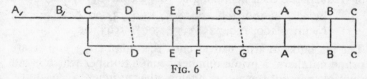

FIG. 6

The displacement of D and G clearly appears, and is shown by the dotted lines. The major and minor tones occur alternately in the minor scale, as in the major scale, except at the octave. The ratios to A, of the intervals of the minor scale are shown in the following diagram in which we also show the major and minor tones by signs as before:

If we add harmonies to this scale we find that we must make two of the notes, B, and G, mutable. Consider the chord:

D as the subdominant we have already found, and G is clearly in its right place in the top line of Fig. 6 as being a fourth above it. B should therefore be a minor third below D and minor sixth below G. A minor third contains a major tone and a semitone. B, must therefore be moved down a comma if it is used as a concord with D and G. The ratio GB, is then $\frac{16}{9} \div \frac{10}{9}$, or $\frac{8}{5}$, which is the ratio of the true minor sixth, for example, the ratio of the interval cE, or the interval FA,. But if we wish to form a minor triad on E:

we find that G must be moved up a comma. E is the dominant and we have found its position. B, in its correct place for insertion in the diagram in Fig. 6, would be an octave above B,, and the interval BA, would therefore have a ratio $\frac{9}{4}$. The interval EA, has a ratio $\frac{3}{2}$. Hence the interval BE has a ratio $\frac{9}{4} \div \frac{3}{2}$, or $\frac{3}{2}$, and is a perfect fifth. Or we can observe that the interval from E to B in Fig. 6 is made up of two major tones, a minor tone, and a semitone. To become a concord with E and B, G must therefore be raised by a comma: as is evident indeed from Fig. 6.

These mutable notes present no difficulty for unaccompanied voices or strings. The performers have only to sing or play in tune and the notes so sung or played will find their own positions. This was equally true in the sixteenth century before composers made use of modulation as we know it. But these mutable notes presented then, as they do to-day, a problem for keyboard instruments, a problem to be solved as far as may be only by a compromise. This led to the tuning of virginals and organs in the sixteenth century in what is known as mean-tone temperament, which will be explained in the next chapter. Not for another two hundred years or more was this method of tuning keyboard instruments finally superseded by tuning in equal temperament to meet the demand of later composers for completely free modulation.

The Simple Use of Logarithms.

In calculating the intervals of the pure scale we had to deal only with easy fractions. When we come, in the next chapter, to deal with the intervals of tempered tunings, we shall encounter calculations which involve much more complicated arithmetic, so complicated, indeed, that we shall have to use a labour-saving device invented by the mathematician. This device he calls the use of logarithms; and it is described in the following paragraphs.

There are many to whom detailed arithmetical calculations are so uncongenial that they serve to obscure the conclusions instead of making them clear. These will be inclined to lay the book down at this point. That would be unfortunate, for their requirements have been borne in mind in its writing. They will find in Chapter II that the errors introduced into the scale by temperaments, errors calculated there by means of logarithms, have been turned into fractions of a comma for their use. They will find that they can omit the calculations if they are content to accept the results. In Chapters IV and V they will find it necessary to know the actual frequencies of the notes of equal temperament if they are to see exactly how much damage is done to the pure beauty of the sound by substituting tuning in equal temperament for the pure scale. The frequencies are there calculated by logarithms. But readers can turn instead to Table IV, Appendix I, where the frequencies of equal temperament are set out; while in Table III, Appendix I, they will find the logarithms from which these frequencies are obtained. They should, however, examine the graphical representations of the pure scale on p. 19.

The use we shall make of logarithms is so very elementary that those whose recollection of their use may have become hazy, or to whom they are so far unknown, may well be invited to learn to understand them. They will then derive a certain satisfaction from following in detail the arithmetical calculations, and this will help to give them a clear grip of the results. For their benefit, as well as for those who are accustomed to the use of logarithms, the calculations arising in the next two chapters are therefore given in detail. If they are found too difficult they can be left for a second reading, as explained in the preceding paragraph.

The rule of algebra on which logarithms depend is that which tells us that $x^2 \times x^3 = x^5$. As explained on p. 4, a mathematician uses the letter x only as a kind of shorthand for any particular number we chose. The rule of multiplication we have quoted follows inevitably from the meaning of x^2 and x^3; for x^2 only means two x's multiplied together, while x^3 means three x's multiplied together. We can state the rule more generally as

$$x^m \times x^n = x^{m+n}$$

where again m and n are just the mathematician's names for any numbers we like to choose. Moreover, the mathematician says that to him this has a definite meaning whether m and n be whole numbers or decimals. If x^m was in fact the value of some number, say A, and x^n was the value of another number, say B, we could multiply A and B together by adding m and n if we could find out what number x^{m+n} was. Mathematicians have calculated the values of m and n for all the numbers they are likely to want. The results are printed in tables they use, for which purpose x is given the value 10. Now suppose A is 2; the mathematician says it equals $10^{0.30103}$; and if B is 3, the mathematician says it equals $10^{0.47712}$. He calls 0·30103 the logarithm of 2, or more briefly log 2; and he calls 0·47712 the logarithm of 3, or log 3. And since $A \times B$ was x^{m+n}, and m has become 0·30103 and n 0·47712, he says that the product of 2 and 3 is a number whose logarithm is 0·30103+0·47712, or 0·77815; and he finds from his tables that this decimal is log 6.

In reading the preceding paragraph the non-mathematical reader may have begun to wonder what 0·30103 tens multiplied together can possibly mean. The mathematician does not ask us to consider what it means. All he asks us to do is to agree that

$10^{0.30103}$ shall obey the same rule as 10^2 and 10^3. If we agree to this he will deduce practical results. He did so, for example, in designing a slide-rule. The scale used in making a slide-rule is, in fact, divided into logarithms. When we add the readings on the two parts of the slide-rule in order to multiply we are adding two logarithms, in other words, carrying out the mathematician's instruction to agree that $x^m \times x^n = x^{m+n}$ even if m and n be decimals. We know by experience that it is safe to trust his slide-rule and that it gives us practical results. It is just as safe to trust his rule of multiplication, and it will give us results just as practical.

To explain how to use logarithms we have chosen simple numbers like 2 and 3 which any one can multiply without using logarithms. But logarithms are used in precisely the same way for numbers that we cannot multiply so readily. To use them we have two simple rules: to multiply two numbers add their logarithms; to divide one by the other subtract their logarithms.

This is all we need to know except that, as stated, log 2 is 0.30103 to five places of decimals and log 3 is 0.47712 to five places of decimals. For if we want to find log 8 we say it equals log $(2 \times 2 \times 2)$ or (log 2 + log 2 + log 2) or 3 log 2 or 0.90309.

Similarly, if we want to find log 5 we know it is log $\dfrac{10}{2}$ or log

10 minus log 2. Since 10^1 is 10 we know that log 10 must be 1. Therefore log 5 is 1 − 0.30103, or 0.69897. Of course, if we had tables we could look up log 5, when we should find the same result. But those who have no tables of logarithms will find they can manage quite well if they know log 2 and log 3. As illustrating the rules for the use of logarithms the calculations in the following two chapters are made with these two logarithms alone. But in Appendix I there is given a table of the logarithms of a few numbers some of which are useful in calculating scales and temperaments. These can be used, if desired, to simplify the arithmetic.

The great advantage of calculating the logarithms of the ratios of intervals, which we will call the logarithms of the intervals, is that they enable us to compare the sizes of two intervals and to represent intervals graphically. Thus, to find how much larger an interval of a major tone, say, is than an interval of a minor tone, we divide the logarithm of a major tone by the logarithm of a minor tone.

We can illustrate the truth of this statement by a diagram in which we compare directly an octave as represented by the actual frequencies of the notes and an octave as represented by the logarithms of the intervals. To add two intervals we have to multiply their ratios, but we add the logarithms of their ratios. We can therefore draw the logarithms of the intervals to scale and simply pile them up in a column like bricks. On the other

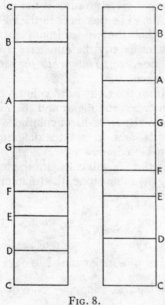

FIG. 8.

hand, to calculate the frequencies of the notes of the scale we must begin with the frequency of the bottom note. For this purpose we will consider the notes in the octave beginning with middle C of the musical clefs and take the frequency of middle C as 256. If we multiply 256 in turn by the fractions which appear in Fig. 2 we shall find the frequencies of D, E, F, &c., at the bottom of the treble clef. For example, the frequency of E is $256 \times \frac{5}{4}$, or 320. On the left of Fig. 8, above, the frequencies are drawn to a scale which has 256 units in the column. On the right the logarithms of the intervals are drawn to a scale which has 301 units (representing the first three digits of log 2) in a column of equal length.

In the left-hand column the distance between C and D is the

same as the distance between B and c. For if the frequency of C is 256, that of D is $256 \times \frac{9}{8}$, or 288, and the difference is 32 vibrations. The frequency of B is $256 \times \frac{15}{8}$, or 480, while that of c is 256×2, or 512, again a difference of 32 vibrations. The left-hand column gives no idea of the relative sizes of the intervals of the scale. The distances in the right-hand column, on the other hand, show the two semitones exactly the same size, the three major tones exactly the same size, and the two minor tones exactly the same size. *Thus, while the ratio fixes the interval, the logarithm measures it.*

It will be seen from the right-hand column how small in fact is the difference between the major and the minor tones. The figures used to draw the right-hand column will be found in a table in Appendix II, and, in fact, the diagrams in Figs. 2 to 7 have been drawn to scale in this way.

For future reference we will calculate the logarithmic values of the major tone, the minor tone, the (diatonic) semitone, and the comma.

$$\log \frac{9}{8} = \log \left(\frac{3 \times 3}{2 \times 2 \times 2} \right) = 2 \log 3 - 3 \log 2$$
$$= 0 \cdot 95424 - 0 \cdot 90309$$

Hence the logarithm of a major tone is 0·05115.

$$\log \frac{10}{9} = \log 10 - 2 \log 3 = 1 - 0 \cdot 95424$$

Hence the logarithm of a minor tone is 0·04576.

$$\log \frac{16}{15} = \log \frac{2^5}{30}$$
$$= 5 \log 2 - \log 3 - \log 10$$
$$= 1 \cdot 50515 - 1 \cdot 47712$$

Hence the logarithm of a (diatonic) semitone is 0·02803.

$$\log \frac{81}{80} = \log \frac{3^4}{2^3 \times 10}$$
$$= 4 \log 3 - 3 \log 2 - \log 10$$
$$= 1 \cdot 90848 - 1 \cdot 90309$$

Hence the logarithm of a comma is 0·00539, which in turn is the difference between 0·05115 and 0·04576, as it should be.

Finally, the number of commas in a major tone

$$= \frac{0.05115}{0.00539} = 9\tfrac{1}{2} \text{ nearly.}$$

The last calculation shows the usefulness of logarithms for comparing the sizes of two intervals, that is, for measuring intervals. We cannot compare the sizes of two intervals by dividing the ratio of the one by the ratio of the other. We know that if we do that we obtain as an answer, not the relative size of the two intervals but something quite different, the ratio of the interval which is the difference of the two intervals with which we started. Thus, had we divided the ratio of a major tone by the ratio of a comma, we should expect to find the ratio of the interval by which a major tone exceeds a comma. That interval is a minor tone. This, in fact, is what we do find. The ratio of a major tone is $\frac{9}{8}$; the ratio of a comma is $\frac{81}{80}$. If we divide the first by the second we have $\frac{9}{8} \div \frac{81}{80}$, or $\frac{9}{8} \times \frac{80}{81}$, or $\frac{10}{9}$. But when we divide the logarithm of a major tone by the logarithm of a comma we obtain quite a different answer, namely, $9\tfrac{1}{2}$. That answer is the number of commas in a major tone. This is the logical consequence of the fact that, as we have already seen, if we wish to add intervals we add their logarithms, which measure them; in other words, if our bricks had been the logarithms of commas, nine and a half of them piled up would produce a column representing the logarithm of a major tone.

The relative sizes of a major tone, a minor tone, and a semitone are:

$$1 : \frac{0.04576}{0.05115} : \frac{0.02803}{0.05115}$$

$$\text{or } 1 : 0.89 \qquad : 0.55$$

A semitone is more than half a major tone and *a fortiori* more than half a minor tone. Or we could calculate the sizes of the major tone, the minor tone, and the semitone in relation to the mean of the sizes of the major and minor tone which would then be taken as 1. If we did so these ratios would become:

$$1.05 : 0.94 : 0.58$$

and this way of showing them will be useful for comparison with the intervals of mean-tone temperament when we come to Chapter II (p. 28).

We have supposed the frequency of middle C to be 256. In

fact, the lowest modern pitch, at normal temperatures, gives the frequency of middle C in equal temperament as 258·7.[1] The difference is very slight and pitches have varied a good deal in the past. If we suppose a string to vibrate once a second it would give an inaudible sound. But a note eight octaves higher would have a frequency 2^8, which is 256. If we made this middle C, the bottom C on the organ, using an 8-foot stop, would then be six octaves higher than the inaudible note with a frequency of 1. It would have a frequency of 2^6 or 64. The bottom pedal C, with a 16-foot stop, which sounds the same note as the bottom C on the pianoforte, would therefore have a frequency of 32, and the bottom pedal C on the organ with a 32-foot stop a frequency of 16: which is both useful and easy to remember.

In his treatise Helmholtz used a German pitch which would give to the bottom pedal C with a 16-foot stop a frequency of 33. This made middle C a note with frequency 264. We shall use the frequency 256 in this book, as being very close to modern pitch and very convenient for calculation as well as a good unit with a philosophical basis.

If, having studied in this Chapter the construction of the pure scale from the concords of the polyphonic period, the reader is curious to learn of the earlier, melodic, development of the European scale he will find an interesting account of what is known of it from Greek times onwards in Chapter II of Parry's *Art of Music*. In this he will also find descriptions of melodic scales developed in other parts of the world which bear little or no resemblance to the European scale; in at least one of which the fifth, harmonically so important to us as the dominant, is not to be found.

---

[1] This pitch is known as diapason normal and corresponds to a frequency of 435 for a′. Diapason normal appears to have been adapted for concert use in this country by taking the interval a′c″ in just intonation. When tuned in equal temperament to the c″ so obtained, the organ gives the strings an a′ to tune by which is slightly sharp on the a′ of diapason normal.

II

TEMPERAMENTS

THE study of the pure scale in the opening chapter was prefaced by a quotation from Parry which made it clear at the outset that musical scales are created and developed by the development of the art of musical composition. Before the golden age of polyphony, musical art attained a stage which demanded a high degree of vocal skill in the use of plainsong, as Terry tells us in his essay on Palestrina in *The Heritage of Music*. To attempt to discuss the melodic practice of these early periods or to trace the story back to the Greek modes would be to travel far outside the scope of this book: such matters would be more appropriate for a treatise on the history of music. But clearly it was fortunate for European music that the melodic modes of its early history should have lent themselves so readily to the singing of concurrent melody which became counterpoint. European music owes a great debt to the perception of consonance by the Greeks and its use in the construction of their modes; for, as we shall see in later chapters, there is an acoustical quality in consonance which was accepted in the main by the art of music as the basis of 'concord' in the polyphonic period. That the essential nature of consonance was then a mystery is beside the point: its application to their art was the product of the musical insight and the sensitive ear of composers. The result, in the hands of the great masters, was music which for pure beauty of sound has never since been approached.

It is also true that the modes, embodying as they did a delicate differentiation of melodic colour, were not scales as we know them; for the sense of tonality was still to come. On the other hand, the scales examined in Chapter I were the pure major and minor scales with an established tonic, dominant, and subdominant. But, essentially, the other ecclesiastical modes differed from the major mode only in the order of their intervals.

In all contrapuntal writing the intervals themselves, for harmonic reasons, would be the intervals of the pure scale, or just intonation as we may now call it, namely, the major tone, the minor tone, and the true diatonic semitone. Moreover, as the editors of *Tudor Church Music* point out in the fascinating introduction to Volume I of the Carnegie Trust Edition, even in Taverner's time in the first quarter of the sixteenth century, although the outlook of the composer in the matter of texture and combinations of rhythm was still horizontal, 'his mind was at last —what it had never been before—chained and tethered to the perpendicular chord as seen from the bass upwards,' while the whole musical sense of Europe was moving from the modal system of medieval plainsong to the system of major and minor. As R. O. Morris observes in his authoritative work on *Contrapuntal Technique in the Sixteenth Century*, this movement of musical sense is particularly evident in the music of our own country.

The seventeenth century saw a musical revolution in which composers established a claim to much greater freedom than had been permitted by the practice of the polyphonic period. Tovey explains, in his article on harmony in the *Encyclopædia Britannica*, that discords which had required preparation were now accepted as essential,[1] and that the aim of the successors of Monteverdi was to establish an orientation of the major and minor keys through a system of relations between tonic, dominant, and subdominant harmonies, which could employ essential discords and allow free modulation of key.

It was shown in the first chapter that to effect the simplest modulations not only are new notes, sharps or flats, required in the scale, but certain notes require adjustment by a comma to maintain just intonation. The new direction given to musical composition in the seventeenth century was bound to affect the musical scale. Equally the tuning of keyboard instruments in what is known as mean-tone temperament,[2] adopted in the sixteenth century for tuning the virginal and the organ, was to prove inadequate: indeed the tuning of keyboard instruments remained in a state of some uncertainty for another century or more. So far as he himself was concerned, the issue was settled for the clavichord by John Sebastian Bach when he produced his forty-eight preludes and fugues for the 'well-tempered clavier'; but though his practice was eventually followed in the tuning of the pianoforte, the old method of tuning the organ by mean-tone temperament survived in some churches until comparatively recently and was abandoned reluctantly to meet

[1] Readers who have not studied counterpoint will find in Sir Donald Tovey's article a description of what the musician calls 'unessential notes', such as passing notes, suspensions, or decorating notes which are not part of the 'essential' harmony.

[2] This method of tuning is said to have been invented by a Spanish musician, Salinas (1513–90). In his day it would solve for a keyboard instrument the problem of the mutable notes, which would occur in the modes just as they do in the major and minor scales and for the same reason.

the demand for a more extensive range of keys and for chromatic usages. In this way the development of the art of music necessitated some departures from the pure scale of the polyphonic period, which is still the natural scale for unaccompanied voices or strings, and eventually substituted equal temperament for mean-tone temperament as a method of tuning keyed instruments. Observe that temperaments are merely methods of tuning keyed instruments: they should not be thought of as new musical scales.

It is important to realize that the development of our modern scale system is part of the history of musical composition. Exactly what this means to the ear of the musician, expressed in scientific terms, we shall learn in Chapter V. It is unlikely that scientific knowledge will influence possible future development of the scale system any more than knowledge of acoustics can have influenced its developments at the hands of composers in the seventeenth and eighteenth centuries, when the science was still in its infancy. The student cannot recall too often the statement that scales are made by writing music, and music is made by composers and not by science. Indeed, we find the great composers of the classical period content to accept the modern scale system, with its acoustical imperfections, in order to secure that development of the art of music which their genius demanded.

To-day, universal familiarity with the piano makes it somewhat difficult to realize that it is tuned in what is only one of several possible ways, one moreover which sacrifices certain perfections of other tunings in order to secure certain compensating advantages. It has been seen that it is not possible to combine both just intonation and free modulation in a single scale or cycle of seven notes to which are added such sharps or flats as are required for 'sharp' or 'flat' keys. Much ingenuity was expended in the last century in devising cycles for insertion in the octave which required many more notes, and additional keys for insertion in keyed instruments; but all this ingenuity failed through the essential inconvenience of the result even when it was not based on unsound principles, as it usually was. The search for the best temperament, or perhaps it would be better to say the temperament of least badness, is a search for the compromise which secures as far as possible the combination of three incompatible requirements: true intervals, free modulation, and convenience, all needed to express on keyed instruments the scale system the composer thinks in.

It will at once be obvious that to cope with the mutable notes which we discovered in Chapter I, or to secure freedom of modulation, the first step is to eliminate from keyed instruments the modifications called for by the difference between the major tone and the minor tone. This is achieved in the only temperaments which have had any real vogue, the mean-tone temperament and the equal temperament, by substituting one single and uniform tone for both the major tone and the minor tone.

Mean-Tone Temperament.

In this temperament the major third is tuned true; and all the tones are made equal to the mean of a major tone and a minor tone. The mean tone must therefore be half a comma less than a major tone and half a comma greater than a minor tone. All the major thirds, which are their sum, are therefore perfect. To add two mean tones we must multiply their ratios, when we shall obtain the ratio of a perfect major third $\frac{5}{4}$. It follows that the ratio of a mean tone is $\sqrt{\frac{5}{4}}$. But to add two mean tones we add their logarithms. It follows that the logarithm of a mean tone is half the logarithm of a major third or $\frac{1}{2} \log \frac{5}{4}$.

It remains to determine the value of a semitone in this temperament. There will be two semitones in the octave of the scale and these are made equal. The octave, regarded as built up of intervals in just intonation, contains two major thirds (C to E and F to A), two semitones of just intonation, and one major tone (A to B). If for each major third we substitute two mean tones and for the major tone we substitute a mean tone plus half a comma we shall have an octave made up of five mean tones, half a comma, and two semitones of just intonation. Dividing the half comma, and giving to each semitone a quarter of a comma in consequence, we obtain an octave of five mean tones and two mean-tone semitones; and each mean-tone semitone equals a semitone of just intonation plus a quarter of a comma.

The fifth and consequently the fourth, which is its complement in the octave, are faulty. A true fifth contains a major third, a semitone of just intonation or a true diatonic semitone, and a major tone. Substituting mean-tone intervals this gives a major third, a mean-tone semitone minus a quarter of a comma, and a mean tone plus half a comma. Thus a true fifth equals three mean tones and a mean-tone semitone plus a quarter of a comma, that is, a mean-tone fifth plus a quarter of a comma.

A mean-tone fifth is therefore too small by a quarter of a comma. It follows that a mean-tone fourth, to complete the octave, exceeds a true fourth by a quarter of a comma.

The next note above the fifth is the sixth, which in just intonation exceeds the perfect fifth by a minor tone. Substituting a mean-tone fifth, which is a quarter of a comma too small, and adding to it a mean tone which is half a comma larger than a minor tone, we obtain a mean-tone sixth which exceeds a true major sixth by a quarter of a comma. In turn, the seventh of the scale, or the leading note, is tuned too flat by a quarter of a comma; but the defect is absorbed into the mean-tone semitone which completes the octave and is, as has been shown, a quarter of a comma in excess of a true diatonic semitone.

We may now set out the intervals produced by mean-tone tuning in two columns showing the extent to which successive notes of just intonation are tempered:

1st degree		o
2nd	,,	$-\frac{1}{2}$ comma
3rd	,,	o
4th	,,	$+\frac{1}{4}$ comma
5th	,,	$-\frac{1}{4}$ comma
6th	,,	$+\frac{1}{4}$ comma
7th	,,	$-\frac{1}{4}$ comma
8th	,,	o

An interesting check on the value of a mean-tone fifth is afforded by the correction which, as we have seen, Didymus applied to the third obtained by Pythagoras through tunings in perfect fifths. An interval of four successive perfect fifths exceeds by a comma the interval of two octaves and a perfect major third. If, for example, we started from C, the fifths C to G, G to d, and a to e′ would be perfect in just intonation. They would all have the ratio 2 : 3. But the interval from d to a would be a comma short of a perfect fifth (see p. 12). If for this interval a perfect fifth were substituted there would be produced an e′ which was a comma sharp. But the interval made up of two octaves and a perfect third remains unchanged when mean-tone tuning is used, and must be the sum of four mean-tone fifths. Four perfect fifths consequently exceed four mean-tone fifths by a comma. Hence a mean-tone fifth is equal to a perfect fifth less a quarter of a comma.

To compare the sizes of the intervals of mean-tone temperament with those of just intonation we may have recourse to logarithms. It has already been shown that the logarithm of a mean tone is equal to $\frac{1}{2} \log \frac{5}{4}$, i.e. $\frac{1}{2} \log \frac{10}{8}$.

Hence, log of a mean tone $= \frac{1}{2} (\log 10 - \log 8)$
$= \frac{1}{2} (1 - 0.90309)$
$= 0.04845(5)$

A mean-tone semitone is made up of a true diatonic semitone and a quarter of a comma. A comma, in turn, is made up of four intervals each equal to a quarter of a comma. Remembering that to add intervals we add their logarithms, it is evident that if we were to write down the logarithm of a quarter-comma four times and add what we had written we must obtain the logarithm of a comma. It clearly follows that the logarithm of one-quarter of a comma must be equal to a quarter of the logarithm of a comma or $\frac{1}{4}$ (0.00539), for it was shown in Chapter I that the logarithm of a comma was 0.00539. It was also shown that the logarithm of a semitone was 0.02803.

Hence, log of a mean-tone semitone
$$= 0.02803 + 0.00134(7)$$
$$= 0.02938$$

The relative sizes of a mean tone and a mean-tone semitone are therefore given by the ratio 1 : 0.60, which may be compared with corresponding ratios found in Chapter I for the intervals of just intonation (see p. 21).

Equal Temperament.

In the equal temperament all the tones are made equal and exactly twice the semitone. By this compromise modulation is made completely free and the maximum of convenience is secured by the sacrifice of some true intervals, and particularly that of the major third.

The octave contains twelve tempered semitones, and is the sum of six tempered tones. To add intervals we multiply their ratios. If therefore we were to write down the ratio of a tempered semitone twelve times and multiply together what we had written we should obtain 2 as an answer. Consequently the ratio of a tempered semitone must be $\sqrt[12]{2}$. Similarly the ratio of a tempered tone is $\sqrt[6]{2}$. This presents a complicated calculation in arithmetic. But to add intervals we add their logarithms. If therefore we were to write down the logarithm of a tempered semitone twelve times, and add, we should obtain the logarithm of an octave, which is log 2 or 0·30103. The logarithm of one tempered semitone is therefore $\dfrac{0·30103}{12}$, which equals 0·02509.

Similarly the logarithm of a tempered tone is $\dfrac{0·30103}{6}$, or 0·05017.

The logarithm of a comma was found to be 0·00539. If we were to write down, eleven times, the logarithm of an interval which was $\dfrac{1}{11}$ of a comma, and add, we should obtain the logarithm of a comma. Hence the logarithm of $\dfrac{1}{11}$ of a comma is $\dfrac{0·00539}{11}$, or 0·00049. This happens to be a logarithm which is needed in translating into fractions of a comma the errors of the tempered scale; so it is useful to note it. As their logarithms show, one-eleventh of a comma is approximately equal to a fiftieth part of a semitone of equal temperament.

The tempered fifth contains seven tempered semitones, and its logarithm is therefore seven times the logarithm of a tempered semitone, or $\left(7 \times \dfrac{0·30103}{12}\right)$, which is 0·17560. The logarithm of a true fifth is (log 3 — log 2), or 0·17609. The tempered fifth is too small and the logarithm of the interval between the true and the tempered fifth is 0·00049. This is the logarithm of $\dfrac{1}{11}$ of a comma. Consequently the defect of a tempered fifth is $\dfrac{1}{11}$ of a comma. It follows that the tempered fourth exceeds the true fourth by $\dfrac{1}{11}$ of a comma.

The tempered third contains 4 tempered semitones, and its logarithm is therefore $\dfrac{4}{12} \times 0·30103$, or 0·10034. The logarithm of a true major third is log $\dfrac{10}{8}$, or (1 — 0·90309), or 0·09691. Thus a tempered major third exceeds a perfect major third by an interval whose logarithm is (0·10034 — 0·09691), or 0·00343, or 7(0·00049). This interval is therefore $\dfrac{7}{11}$ of a comma.

The student is recommended to calculate all the intervals of the tempered scale in this way, when he will find that the logarithm of a tempered minor third is 0·07526 and that a tempered minor third is consequently smaller than a true minor third by an interval equal to $\frac{8}{11}$ of a comma. This is to be expected, for a fifth being the sum of a major and a minor third, the two fractions $+\frac{7}{11}$ and $-\frac{8}{11}$ of a comma give the defect in the tempered fifth. Similar calculations will show that the tempered tone is $\frac{2}{11}$ of a comma less than a true major tone, and $\frac{9}{11}$ of a comma greater than a minor tone; as it should be, of course, since the difference between a major and a minor tone is one comma. The defect in the tempered semitone is an interval whose logarithm is (0·02803−0·02509) or 0·00294, which equals 6 (0·00049). This is obviously the logarithm of an interval equal to $\frac{6}{11}$ of a comma.

The true major sixth is made up of a true fifth and a minor tone. The tempered major sixth is made up of a tempered fifth and a tempered tone. Its error will be the algebraic sum of the defect in the tempered fifth and the amount by which a tempered tone exceeds a minor tone: or ($-\frac{1}{11}$ of a comma $+\frac{9}{11}$ of a comma); which gives $+\frac{8}{11}$ of a comma. The true major seventh is the sum of a true major sixth and a major tone. The error in the tempered major seventh is the algebraic sum of the errors in the tempered sixth and a tempered tone, or ($+\frac{8}{11}-\frac{2}{11}$ of a comma); which gives $\frac{6}{11}$ of a comma. This excess is absorbed in the interval between the leading note and the octave, whose defect is $\frac{6}{11}$ of a comma, as has been seen.

Consequently the intervals produced by tuning in equal temperament can be set out in two columns showing the extent to which successive notes of the scale are tempered.

1st degree	o
2nd ,,	$-\dfrac{2}{11}$ comma
3rd ,,	$+\dfrac{7}{11}$ comma
4th ,,	$+\dfrac{1}{11}$ comma
5th ,,	$-\dfrac{1}{11}$ comma
6th ,,	$+\dfrac{8}{11}$ comma
7th ,,	$+\dfrac{6}{11}$ comma
Octave	o

The fifth and fourth are nearly correct, but the major and minor thirds are faulty. The minor sixth is the difference between the major third and the octave. Its defect is therefore $\dfrac{7}{11}$ of a comma, while the major sixth is $\dfrac{8}{11}$ of a comma too big. The sixths are, of course, the complements of the thirds and must exhibit errors corresponding to those of the thirds.

It is interesting to calculate the error in the fifth by a method suggested by the Pythagorean tuning of the scale by perfect fifths. On the piano the sum of twelve fifths, starting from a given note, is a note seven octaves above the given note. Thus twelve tempered fifths added together give an interval equal to seven octaves. In other words, if we were to write down the logarithm of a tempered fifth twelve times, and add, the answer would be the logarithm of an interval of seven octaves, or seven times the logarithm of an octave, or 2·10721. Consequently the logarithm of one tempered fifth would be $\dfrac{2·10721}{12}$ or 0·17560, which is the value found for it by the previous calculation.

In turn, by using the tempered fifth to tune our scale we can check the results of our calculations. If we start with C, a tempered fifth would give us a G which was $\dfrac{1}{11}$ of a comma flat.

Another tempered fifth would give us a D which was $\dfrac{2}{11}$ of a comma flat. For the interval from D to A in just intonation we want a true fifth less a comma, that is, a tempered fifth less $\frac{10}{11}$ of a comma. If we made the interval from D to A a tempered fifth instead, we should obtain an A which was $\left(\frac{10}{11} - \frac{2}{11}\right)$, or $\frac{8}{11}$, of a comma sharp. Another tempered fifth would give us an E which was $\frac{7}{11}$ of a comma sharp, while B in turn would be $\frac{6}{11}$ of a comma sharp. This, in effect, is what the tuner of a keyboard instrument does.

The Pythagorean tuning by perfect fifths is used to-day in tuning the violin. When a viola and a violin are tuned to the same A, the C string of the viola produces with the E string of the violin an interval equivalent to the sum of two octaves and a Pythagorean third, that is, a just major third plus a comma. It is therefore interesting to observe that string players prefer to avoid the use of open strings, which produce a tone slightly different from that of stopped ones; thus their intonation remains free.

When this method of tuning was used with a longer succession of fifths tuned upwards, octaves tuned downwards were interpolated between every second or third fifth to restrict the compass of the notes used. In this way six of the seven notes of the scale could be obtained. The remaining note, the fourth, would be given by a fifth tuned downwards from the octave. But we should obtain Pythagorean thirds, instead of perfect major thirds, for CE, FA, and GB: E, A, and B would each be a comma sharp. As a consequence the semitones would each be a comma too small, giving the Pythagorean *limma* or Greek hemitone. Pythagoras was quite aware that the sum of twelve perfect fifths was more than seven octaves, the difference being known as the comma of Pythagoras. The comma of Pythagoras was slightly larger than the comma with which we are familiar, the comma of Didymus, for, as it is eliminated in equal temperament by the cumulative errors of twelve tempered fifths, the comma of Pythagoras obviously equals $\frac{12}{11}$ of the comma of Didymus.

We saw in Chapter I that, for reasons founded in Arithmetic, successive tunings in perfect fifths will never coincide again with successive tunings in octaves which start from the same note.

That is why D has to be a mutable note in the pure scale, and why other notes have to be adjusted by a comma when we modulate. For the same reason when, without modulating, we use the chromatic seventh on the supertonic in the pure scale we must be prepared to adjust A. At the end of the next chapter we shall encounter a similar difficulty which arises if we tune in successive perfect major thirds. To quote Parry again: 'An ideally tuned scale is as much of a dream as the philosopher's stone.' He goes on to observe that a scale system may fairly be tested by what can be done with it, and that the scale which afforded Bach, Beethoven, Schubert, Wagner, and Brahms ample opportunities to produce the works they did, is as perfect as their musical art required. 'It will probably be a good many centuries before any new system is justified by such a mass of great artistic works as the one which the instincts and efforts of our ancestors have gradually evolved for our advantage.'

If he has carefully worked through the opening chapters of this book the student will be equipped to read critically, if he so desires, a fuller discussion of scales such as that to be found in Helmholtz's work. Many of the refinements which in the past have been suggested for scales for keyed instruments will then appear to him to be monuments of misplaced ingenuity concealing a modicum of musical truth, or even to be based on wrong principles, as Helmholtz points out. He will, in particular, find it interesting to learn why the equal temperament is more especially suitable for the piano, for which or for its predecessors it was originally developed; and why it is less suitable for the organ unless played softly, and least suitable when the mutation stops and reeds are used. The reasons will be given in subsequent chapters of this book.

III

CHROMATIC NOTES, DECORATING NOTES, INTONATION

In Chapter I it was found that a new note, F♯, had to be added to the pure scale for modulation into the key of G, and as it was the leading note of the scale it was a diatonic semitone, ratio $\frac{15}{16}$, below G. It was also found that a new note, B♭, was required for modulation into the key of F, and it was found to be a diatonic semitone, ratio $\frac{16}{15}$, above A. This note is also the minor ninth of an A an octave lower. A similar note, G♭, is needed a diatonic semitone above F to form a minor ninth with an F an octave lower. In learning modulation the student is introduced to what are described as enharmonic changes. Thus in using the chord of the diminished seventh or minor ninth he learns that a G♭ may be changed into an F♯ in passing from one key relationship to another, and that this enharmonic change is actually made on stringed instruments, though on the piano, with its five black keys, the two notes are identical. It will help him to appreciate this if the positions of F♯ and G♭ are determined exactly by their harmonic relationships.

The position of chromatic notes, from a purely mathematical point of view, will depend on the use made of them by the composer as interpreted by the performer. There will be three uses to take into account: first, the position required for true harmonic relations; second, the indeterminate position given to a chromatic note as a leading note in a decoration of a melody played on the violin; third, the position determined by use in a chromatic scale as played on the violin. The term leading note must here include a falling succession, such as A♭ to G, historically the earliest form of leading note (*vide* Chapter II of Parry's *Art of Music*).

Consider first the position for true harmonic relations. F♯ is required as a major third above D. The ratio of its frequency to that of C will be $\frac{9}{8} \times \frac{5}{4}$, or $\frac{45}{32}$. This is the leading note for the key of G, for $\frac{3}{2} \div \frac{16}{15}$ is $\frac{45}{32}$. G♭, on the other hand, is a diatonic semitone above F; for example, as the dominant minor ninth of the key of B♭. The ratio of its frequency to that of C is $\frac{4}{3} \times \frac{16}{15}$, or $\frac{64}{45}$.

The sizes of the intervals between C and each of these notes may easily be compared by calculating their logarithms. Thus $\log (\text{C to F}\sharp) = \log \frac{45}{32} = 0\cdot14806(3)$, while $\log (\text{C to G}\flat) = \log \frac{64}{45} = 0\cdot15296(7)$. The logarithmic difference between these intervals is $0\cdot00490$, which is 10 times the logarithm of $\frac{1}{11}$ of a comma. It follows that G♭ is $\frac{10}{11}$ of a comma above F♯.

The number of commas in a major tone is found by dividing their logarithms, and is therefore $\frac{0\cdot05115}{0\cdot00539}$, which is $9\cdot5$ to one place of decimals. The number of commas in a diatonic semitone is $\frac{0\cdot02803}{0\cdot00539}$, which is $5\cdot2$. We may represent our intervals graphically by using their logarithms, as we did for the pure scale in Chapter I.

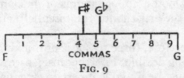

FIG. 9

In this diagram F♯ appears $4\cdot3$ commas above F, and the addition of a semitone or $5\cdot2$ commas brings us to G. The interval between F♯ and G♭ is $0\cdot9$ of a comma.[1]

These exact calculations are interesting, but too much importance must not be attached to them if the notes appear in discords. On physiological grounds, because, as we shall see in Chapter V, dissonant intervals have not the same definition, due to beats between their upper partial tones if they are mistuned, as consonant intervals. On musical grounds, because the

[1] Fig. 9 also exhibits, graphically, the convention employed in musical notation. A diatonic semitone, such as FG♭ or F♯G or E♯F♯, is properly denoted by two different letters using sharps and flats as required. An interval denoted by the same letter using sharps or flats as required, such as CC♯ or D♭D or B♭♭B♭, is a chromatic interval which is not defined by direct harmonic relationship. Such an interval merely measures the chromatic alteration of some note of the diatonic scale of, say, G, or F, or D♭ respectively, required to produce with a neighbouring note a semitone which is not found in that diatonic scale. Theoretically, the ratio of such a chromatic interval in the pure scale may be either 25 : 24 or 135 : 128, as we saw in Chapter I.

melodic line has a strong influence over the quality of the intervals. This is particularly true of decorating notes, which in the technical language of the musician are unessential. There is an instructive passage in the writings of Hauptmann (1792–1868), himself a professional violinist, whose contributions to musical theory are quoted with great respect by Helmholtz. Hauptmann says that if he had to play C♯ as a leading note or D♭ as a minor ninth he would take the first much higher than the second.

'We may assert', he writes, 'that even the mathematically true intonation does not suffice for an animated performance,' and 'an animated intonation' (he is writing of the violin) 'is just as little mathematically true as an animated timekeeping is strictly according to the metronome . . . In two passages C, D♭, C and C, C♯, D, it is certain that the D♭ will be sung flatter than the C♯, although the former mathematically is 15 : 16 and the latter 24 : 25. This is the psychological view of intonation that the clavier can know nothing of.'[1]

This is a clear pronouncement that the melodic character of decorating notes determines their frequency when they are sung or played on the violin. When he is playing decorating notes, such as passing notes, appoggiaturas or grace notes, the dominant feeling of the artist is for the essential note to which they are moving.

We have, therefore, three reasons for thinking of the musical scale as something flexible. First, because, as we saw in Chapter I, certain notes must be mutable if certain essential harmonies are to be in tune. Second, because it is necessary to be able to alter the position of certain notes, for a similar reason, when we modulate. Third, because the positions of decorating notes, which are not part of the essential harmony, are indeterminate. If we assume skilled intonation in playing the violin, the position of notes altered in the first two ways will depend on the accuracy of the artist's ear: the position of the decorating notes will depend, rather, on his musical feeling.

Incidentally it will be noted that by assigning to the interval CC♯ the ratio 24 : 25 Hauptmann must have assumed D to be in the position which it reached in the first example on p. 12; that is, after suffering mutation to make it a perfect fifth below A. It is evident that the difference between C and D will then be 8·5 commas, and that between C♯ and D♭ 1·9 commas, as the student may confirm by working out the logarithmic differences.

[1] The quotation is taken from Pole's *Philosophy of Music*.

There are, in fact, more than two theoretical positions for C♯ and they are separated from each other by intervals of a comma: to distinguish between them Helmholtz adopted, with slight modification, a notation due to Hauptmann which was based on tunings in fifths, a notation which would suggest itself naturally to a violinist whose instrument is tuned in fifths. There is an interesting passage in Helmholtz's work which deals with the distinction between the different positions of C♯ whatever may be their relation with D♭.

'The old attempts to introduce more than 12 degrees into the scale have led to nothing practical, because they did not start from any right principle. They always . . . imagined only that it was necessary to make a difference between C♯ and D♭, or between F♯ and G♭, and so on. But that is not by any means sufficient, and is not even always correct.'

In the same chapter he discusses the effect of equal temperament on the piano, the organ, and the orchestra.

'Orchestral instruments can generally alter their pitch slightly. Bowed instruments are perfectly unfettered as to intonation; wind instruments can be made a little sharper or flatter by blowing with more or less force. They are, indeed, all adapted for equal temperament, but good players have the means of indulging the ear to some extent.'

This will be clear to those who are familiar with the playing of good orchestras. Helmholtz continues:

'Hence, passages in thirds for wind instruments, when executed by indifferent players, often sound desperately false, whereas good performers, with delicate ears, make them sound perfectly well.'

For the orchestral player the only practical meaning of just intonation is 'playing in tune'.

In passing it may be observed that, while he is insistent that the basis of music is melody, when he discusses the application of his discoveries to the art of music Helmholtz is mainly preoccupied with harmonic effect. If he gives insufficient weight to other aspects of musical technique this is, in part, deliberate: in the closing sentence of his final chapter he refers to them, with the intellectual honesty of a great thinker, as matters 'which I should feel myself too much of an amateur' to investigate. Helmholtz accepted as current doctrine the harmonic procedures which he discusses; but he had a just estimate of them:

'I should consider a theory which claimed to have shown that all the laws of modern Thorough Bass were natural necessities, to stand condemned as having proved too much.'

In conclusion, mention should be made of the position of F♯ in a chromatic scale played on the violin. The tendency of the player would be to make the chromatic semitones all equal, and in so far as he succeeded F♯ would find a place in the middle of the diagram on p. 35 and half-way between the positions of F♯ and G♭ as there shown.

In Chapter II it was stated that mean-tone temperament, used within living memory for tuning organs, was abandoned to permit a wider range of keys. The method of tuning was to determine the position of the notes of the scale, including the black notes, by means of successions of fifths flattened by a quarter of a comma, octaves tuned downwards being interpolated as necessary to limit the compass of the notes used. It has been seen that four such fifths would produce a perfect major third. Thus this method of tuning produced correct intervals CE, GB, DF♯, EG♯. But there the trouble began. For the key of which C is the true major third it is necessary to have a true A♭: G♯ will not do at all. It gives, with the C above it, a major third which exceeds a perfect third by 1·9 commas. This may be verified by calculating G♯ as the sum of two perfect thirds, ratio $\frac{25}{16}$ to C, and A♭ as a true minor sixth above C, ratio $\frac{8}{5}$. The difference is an interval with a ratio $\frac{8}{5} \div \frac{25}{16}$, or $\frac{128}{125}$. Its logarithm is 0·01030 or 1·9 times the logarithm of a comma (0·00539). Or by using the results obtained in Chapter II we could verify this without using logarithms. The major third in mean-tone temperament is true; but three major thirds on the pianoforte make an octave. A major third on the pianoforte exceeds a true major third by $\frac{7}{11}$ of a comma. Three of them will exceed three true major thirds by $\frac{21}{11}$ of a comma, or 1·9 commas. The substitution of G♯ for A♭ caused a trying 'wolf'. Consequently organs tuned in the mean-tone temperament could be used only for a limited number of keys.

The next two chapters will deal with acoustical phenomena which are significant in considering how important from the point of view of pure beauty of sound the errors introduced into the scale of keyboard instruments by the adoption of temperaments may be.

IV

COMBINATION TONES

IF two notes whose frequencies are M and N are sounded together *loudly* a tone of frequency $M-N$ may be heard. Among the first to observe this phenomenon was the violinist Tartini (1692–1770), who called it a 'third tone', and made his pupils acquainted with it as a guide to correct intonation in double stopping. On p. 46 are given the positions of this third tone for various intervals. Helmholtz showed that there is likewise produced a tone of frequency $M+N$, which he called a summation tone to distinguish it from Tartini's third tone, which he called the first difference tone.

The cause of this phenomenon was not properly understood until Helmholtz investigated it. The sensation of a musical note was supposed to be due, not to any actual musical vibration in the air of the frequency heard by the ear, but to an effect produced in the ear itself by rapid beating which was mistaken for a musical note outside the ear. As every musician knows, beats which cause an unpleasant sensation are heard when two notes about a semitone apart are sounded together. As the interval is increased the beats become more rapid, and eventually less unpleasant; finally they cease to give any unpleasant feeling or discomfort and are then no longer observed by the ear. It was supposed that when the beats became too rapid to cause any discomfort they produced the sensation of a musical note. The idea seems so reasonable, on arithmetical grounds, that it is still liable to repetition. Helmholtz showed why this explanation is not correct.

In the first place, it provided no explanation of the summation tone. In the second place, Helmholtz showed that two notes sounded loudly on a suitable siren can excite a sympathetic vibration in a membrane tuned to the pitch of the difference tone to be detected: in other words, the difference tone must then exist as an actual musical vibration in the air. Rücker and Edser showed that a similar effect can be produced on a tuning-fork whose frequency is that of the difference tone. In the third place, a general observation, confirmed by experiment, leads to the conclusion that the ear hears motion in the air as a tone of a particular frequency only if a perfectly simple vibration of that frequency actually exists in the air. The nature of the simple vibration referred to will be explained in Chapter VI. Beats, which are due to two such simple vibrations getting in and out of step so to speak, due to their difference of frequency, are not accepted by the ear as a third simple vibration of the same kind. Moreover, it can be shown mathematically that if the two notes of frequency M and N with which we started are sounded *loudly*, other actual vibrations of the simple kind accepted by the ear as a musical tone are bound to be produced in the air, and that their frequencies will be those of the difference and summation tones, known collectively as combination tones.

On the other hand, Helmholtz concluded that there existed in the ear itself certain conditions, due in part to lack of symmetry, which could and did produce an effect giving the sensation of a combination tone; an effect to be distinguished from any effect which rapid beating could cause. In suitable circumstances, if the notes producing it are loud, this effect sounds louder than the sensation actually produced by the vibration of the combination tone in the air.

Helmholtz's conclusions have been the subject of much controversy, but are generally accepted to-day as established. The practical conclusion, which is important in music, is that combination tones are heard if the exciting notes, or generators as he called them, are *loud*. The student can verify this on the organ by sounding a fifth on a loud stop such as the open diapason. A combination tone an octave below the lower note of the fifth will be heard. It is a good plan to sound the lower note of the fifth first and add the upper note of the fifth afterwards. Careful listening will show that the lower octave will sound faintly at the same time. It is less easy to hear combination tones on the piano because with a percussion instrument the intensity of the sound diminishes rapidly. But if a fifth is struck loudly in the treble clef, the lower octave can be heard to sound momentarily: it is more readily detected if, as a guide to the ear, the lower octave is first touched lightly on the keyboard and the finger is then raised before the fifth is struck. Better still, if the fifth $c''g''$ and the fourth $c''f''$ are struck alternately, the first difference tone can be heard, faintly, to leap to and fro between c' and f.

If, therefore, two tones of frequencies M and N are sounded loudly, M being greater than N, there is produced a first difference tone with frequency $M-N$. It is much weaker than the generators. In turn, the first difference tone produces with the tone of frequency N a second difference tone of frequency equal to the difference between N and $M-N$. This is weaker still. There is, of course, another second difference tone produced by the first difference tone and the tone of frequency M. But as it has a frequency $M-(M-N)$, which equals N, it coincides with the lower generator and cannot be distinguished. The summation tone has a frequency $M+N$ and is usually so faint as to be inaudible. Helmholtz shows how it can be made audible on a suitable siren. Other combination tones are so faint that they may be regarded as negligible.

The positions of the difference tones are interesting. First suppose that the generators are less than an octave apart. In that case $2N$, being the frequency of a note an octave above the lower generator, will be greater than M. Subtract N from each of these, and it will be seen that N is greater than $M-N$. In other words, the first difference tone, having a smaller frequency than the lower generator, will be below it in pitch. Moreover, the greater the difference, within the octave, between M and N, the greater becomes $M-N$, and the less, therefore, the difference between N and $M-N$. In other words, the closer becomes the pitch of the first difference tone to that of the lower generator. This fact is exhibited by the series of first difference tones shown on p. 46. The second difference tone in the case of generators less than an octave apart will have a frequency $2N-M$.

Now suppose the generators to be more than an octave apart. The octave above the lower generator will be below the upper generator, and consequently its frequency, which is $2N$, will be less than that of the upper generator which is M. If N be subtracted from both these it will be seen that N is less than $M-N$. That is, the frequency of the lower generator will be less than that of the first difference tone, and consequently its pitch will be lower. In other words, the first difference tone will be between the two generators. The frequency of the second difference tone will be $M-2N$.

If the generators are exactly an octave apart, M and $2N$ will be equal. The first difference tone, $M-N$, will be equal to N. The second difference tone formed with the generator M will, as always, coincide with the lower generator. But the other second difference tone formed with the other generator will have a zero frequency, and will not exist.

Beats due to Difference Tones.

For the present, attention will be confined to generators which have no upper partial tones, that is, which are pure tones. Suppose two generators, each a pure tone, to be sounded together in perfect unison, that is, with the same frequency. Let one of them be gradually raised. This is experimentally possible with suitable apparatus. As the pitch of one of the generators is raised beats are heard, due, as remarked above, to the vibrations getting out of step. The effect may be compared to that which can be heard when two ill-paired horses drawing a vehicle are trotting down a street. The stride of the two horses would differ slightly, so that one would be trotting slightly faster than the other. A beat reaches its maximum intensity every time the vibrations are in step. It reaches its minimum intensity every time they are completely out of step. The rate of the beats is equal to the difference of the frequencies of the generating tones.

So long as the beats are slow they are not unpleasant. When the interval reaches about a semitone for notes in the treble clef, the sensation produced by the beats causes the maximum irritation in the nerves of the ear. As the interval is further increased the sensation becomes less irritating. Finally, the sensation ceases when the generating tones are about a minor third apart.

The irritation caused to the nerves of the ear by beating is to be compared to that caused to the nerves of the eye by a flicker. If light increases and diminishes very slowly, it does not irritate like a flicker, though it is annoying in a way that very slow beats are not annoying to the ear. As the rapidity of the change increases, the flicker becomes intensely irritating. But if it increases still further the flicker becomes less irritating; and eventually, as in a good modern cinema, the eye fails to observe the flicker. But we know that physically the flicker is still there.

The same thing happens to the ear. When the beats become sufficiently rapid they cause no irritation, and the *sensation* of beating ceases. But actually the *physical* cause of the sensation is still there. Sufficiently rapid beats simply do not produce a sensation in the ear, and musically do not exist. That is why the unpleasant quality in the sensation of beating is properly described as a physiological effect in the ear itself. The rate of the physical beat increases steadily the whole time. In the ear it produces a sensation at first not unpleasing, then increasingly irritating, then less and less irritating, and finally the sensation ceases to exist.

To return to our difference tones. It has been seen that the sensation of beating disappears, for notes in the treble clef, when the generators are about a minor third apart. But faint beats are heard when their generators, *if sounded loudly*, are nearly but not exactly a fifth or an octave apart. If the generating tones were produced by tuning-forks on resonance boxes (which can give a nearly pure tone[1]) the beats close to the octave would be strong enough to be used to tune the forks to an exact octave. (It is easy to flatten the pitch of a fork by sticking wax on the ends of its prongs.) But the beats heard at the fifth are much fainter than those heard at the octave.

The beats heard at the octave and the fifth between pure tones are due to difference tones, for the generators have passed beyond the widest interval that produces audible beating. The position of the difference tones may be shown in staff notation as follows, and we will use Helmholtz's device and represent the generators by minims, the first difference tone by crotchets, and the second difference tone by quavers.

FIG. 10

Assume, as before, that middle C, or c′, has a frequency of 256. The frequency of C in the treble clef, or c″, will then be 512. The first difference tone will have a frequency of 256 and will coincide with the lower note of the octave. There will be no audible beats, and only one second difference tone, which will coincide with the first difference tone and with the lower generator. Now suppose the top C to be mistuned and to have a frequency of 508. The first difference tone will have a frequency of 508—256, or 252. This will produce 4 beats a second with the generator at middle C. There will be two second difference tones. One will be 508—252, or 256, and will be masked by the lower generator, as usual. The other will be 256—252 or a note with a frequency of 4, which is outside the range of audibility since it is two octaves below the bottom note on a 32-foot stop on the organ.

Consider now the case of the higher generator at G in the treble clef. Its frequency will be $256 \times \frac{3}{2}$, or 384. The first difference tone will have a frequency of 384—256, or 128. This is C in the bass clef. One second difference tone will have a

[1] See footnote, p. 101.

frequency of 384—128, or 256, and this will, as usual, be masked by the lower generator. The other second difference tone will have a frequency of 256—128, and this will coincide with the first difference tone on the C in the bass clef. Now suppose the fifth to be slightly mistuned and suppose its frequency to be 382. The first difference tone will have a frequency of 382—256, or 126. This is a note very slightly below the C in the bass clef. It produces no audible beats with the generators. There will be two second difference tones. One will have a frequency of 382—126, or 256. As usual, it will coincide with and be masked by the lower generator on middle C. The other will have a frequency of 256—126, or 130. This is a note slightly above the C in the bass clef. The only notes within audible beating distance are the first and second difference tones, the one a slightly flattened C in the bass clef, the other a slightly sharpened C in the bass clef. They will produce 4 beats a second. But these beats will be much fainter than those produced by the mistuned octave. So far as disturbance of the total volume of sound in the ear by beats due to difference tones is concerned, mistuning of the octave will be more unpleasant than mistuning of the fifth.

It will, no doubt, occur to the student that in tempered scales mistuning of the fifth occurs, and must produce an effect on the difference tones if the mistuned fifth is sounded loudly with the tonic. This, in fact, happens. If he has a table of logarithms[1] he can calculate the rates of the beating produced by the difference tones of the mistuned fifths of mean-tone temperament and equal temperament. In equal temperament G is $\frac{1}{11}$ of a comma below its place in just intonation. The logarithm of this interval is 0·00049. It is the logarithm of a ratio between two frequencies, and it is independent of the pitch. The frequency of G is 384 in just intonation. If the frequency of the tempered fifth be x the ratio of the interval is $\frac{384}{x}$ and its logarithm is therefore $\log 384 - \log x$. We thus find that $\log 384 - \log x = 0·00049$

$$\text{or } \log x = \log 384 - 0·00049$$
$$= 2·58433 - 0·00049$$
$$= 2·58384$$
$$= \log 383·57$$

Consequently the frequency of tempered G is 383·57. The first

---

[1] See Table III, Appendix I. N.B. log 384 = 7 log 2 + log 3

difference tone with C (256) has a frequency of 127·57. The second difference tone has a frequency of 128·43. The rate of beating between them is 0·86, or, say 0·9, rather less than one a second. The beats are faint and their rate is not rapid enough to cause any appreciable confusion of tone.

In mean-tone temperament the defect in the fifth is a quarter of a comma. Suppose, as before, that the frequency of G in the treble clef in mean-tone temperament is x. We have found that the logarithm of a quarter of a comma is 0·00134 (see p. 28). As before, it follows that

$$\log x = 2·58433 - 0·00134$$
$$= 2·58299$$
$$= \log 382·8$$

Consequently the frequency of the mean tone G is 382·8. The first difference tone with C (256) has a frequency of 126·8. The second difference tone has a frequency of 129·2. The rate of beating between them is 2·4, or about $2\frac{1}{2}$ beats in a second. The beats are faint and they are not sufficiently rapid to cause any marked confusion of tone, though they are three times as rapid as those between the difference tones of the equal tempered fifth.

In these calculations the generators were supposed to be pure tones. In the next chapter we shall consider the effect when the notes are complex, containing the upper partial tones as well as the fundamental tone. This introduces new factors which cause disturbance of tone through mistuning.

Position of Combination Tones.

For the simpler intervals the position of the first difference tones can readily be ascertained by referring to the position and relative frequencies of the notes of the harmonic series. Thus a fifth is the interval between two notes whose frequencies are in the ratio 3 : 2. The difference between these numbers is 1, and the first difference tone has a frequency in the ratio 1 : 2 with the frequency of the lower generator; that is, it is an octave below it.

In this way we may work out the first difference tones for all the intervals of the pure scale, and represent them, as in Fig. 11 overleaf, in staff notation, using minims, as before, for the generators and crotchets for the first difference tones.

The only note which we cannot readily place is the one given by a difference of 7, because, as was noted at the beginning of

Chapter I, the seventh note of the harmonic series cannot be accurately represented by any note of the stave. To ascertain the exact position of the notes under which an asterisk has been

* The seventh of the harmonic series is flatter than the minor seventh of the fundamental.

FIG. 11

placed in the above diagram, actual frequencies must be used. Taking the frequency of c′ to be 256, the frequency of c″ will be 512 and that of c‴ 1024. The frequency of d″ will be $512 \times \frac{9}{8}$, or 576. The frequency of the first difference tone of c‴ and d″ will be 1024 − 576, or 448.

In the previous chapter it was found that F♯ is lower than G♭; and A♯ is lower than B♭ by an equal amount, for both FG and AB are major tones in just intonation. A♯ is two diatonic semi-tones below C and its frequency at a′♯ is therefore $512 \times \frac{15}{16} \times \frac{15}{16}$ or 450. This is a note just above t⸗e first difference tone of c‴ and d″, or, what is the same thing, the seventh note of the harmonic series on C as a fundamental (for 64×7 is 448). The ratio of the interval by which it exceeds the seventh note of the C harmonic series is $\frac{450}{448}$. To find the logarithm of this interval it is necessary to know that log 7 is 0·84510. Knowing this, it will be a useful exercise for the student to work out the interval $\frac{450}{448}$ in elevenths of a comma. The answer is $\frac{4}{11}$ of a comma. We already know that the interval between B♭ and A♯ is $\frac{10}{11}$ of a comma. This gives an exact idea of the position, below A♯, of the seventh note of the C harmonic series. It is clearly $\frac{14}{11}$ of a comma below B♭, and this is twice the error of the major third in equal temperament. It has been shown as A♯ in the figure; and this has the advantage that not only is it the nearest note we can show in the stave but it emphasises, to any student of music, the discordant character of these difference tones.

In the next chapter we shall learn to distinguish between 'discord', which is a musical term, and 'dissonance', to which Helmholtz gave a physiological meaning, and we shall find that the art of music may treat as discords intervals which, physiologically, are consonances.

The eleventh and thirteenth notes of the harmonic series are not shown quite accurately on the stave. Consider the positions of the first summation tones, shown below by crotchets, as usual. It is obviously fortunate that the summation tones are very faint indeed; otherwise they would disturb the harmony of the major common chord:

The sign ⌣ is used by Helmholtz to show that the note must be flattened a little.

FIG. 12

V

CONSONANCE AND DISSONANCE, MISTUNING, CONCORDANT TRIADS, EFFECT OF TEMPERAMENTS, SOME THEORIES OF HARMONY

In the opening chapter it was stated that intervals formed by two notes whose frequencies are in a simple ratio are musically pleasing. The choice of such intervals is, we know, a matter in which the ear exercises unfettered judgement that owes nothing to any theory. That is why we referred to Pythagoras as 'verifying' or 'standardizing', by the vibrations of strings, a physical relationship for the notes of these intervals. The Greeks were greatly interested in the simple numerical ratios of vibrating strings that gave the intervals which their ears selected; and they began the long series of conjectures about a natural basis for music to which reference was made in the Introduction.

It was Helmholtz who first provided a firm basis for the consideration of this question of 'consonance', as it is called; and at the beginning of his classical treatise there is an interesting summary of earlier attempts to provide an answer.

'This relation of whole numbers to musical consonances was from all time looked upon as a wonderful mystery of deep significance. The Pythagoreans themselves made use of it in their speculations on the harmony of the spheres. From that time it remained partly the goal and partly the starting point of the strangest and most venturesome, fantastic or philosophic combinations, till in modern times the majority of investigators adopted the notion accepted by Euler' (1707–83) 'himself, that the human mind had a peculiar pleasure in simple ratios, because it could better understand them and comprehend their bearings. But it remained uninvestigated how the mind of a listener not versed in physics, who perhaps was not even aware that musical tones depended on periodical vibrations, contrived to recognize and compare these ratios of the pitch numbers. To show what processes, taking place in the ear, render sensible the difference between consonance and dissonance, will be one of the principal problems in the second part of this work.'

In brief, Helmholtz proceeds to find out how the ear is able to say whether two notes sounded together are 'in tune'.

To attempt to summarize the chapter in Helmholtz's work which deals with the anatomy of the ear and physiological functions he assigned to its nervous appendages would take us too far afield. But we may indicate the nature of his conclusions by an analogy which he himself used. Every musician knows that an object which can vibrate freely at a certain frequency may be excited to emit a musical note by sympathetic vibration induced by a loud note of that frequency. The story is told of a famous singer whose powerful voice could excite in a wine-glass sympathetic vibrations of such violence as to shatter the glass. If the loud pedal of a piano is pressed down, so as to raise all the dampers and leave all the strings free to vibrate, a note sung over the strings will cause the instrument to sound the same note by sympathetic vibration of the strings affected. Helmholtz conceived the nervous appendages in the ear—known to physiologists by the name of Corti, who discovered their detailed structure in 1851—as being set into sympathetic vibration by sounds heard by the ear. It is as if each string of the piano were connected to the brain of the listener by a wire along which electric currents could pass if the string vibrated. The analogy gives a clear picture, but it must not be pressed too far.[1] The appendages in the ear can vibrate in such a way as to distinguish vibrations separated, in the middle registers, by much smaller intervals than a semitone; while in the extreme bass, and with notes in the treble beyond the range of the pianoforte keyboard, the ear is unable to say what are the pitches of notes separated by more than a semitone. We will return to this in Chapter VII. Moreover, the analogy ignores any part played by the soundboard.

Now when the same, or narrowly separated, parts of the nervous appendages of the ear are excited to make sympathetic vibrations by two musical notes close together the ear experiences the sensation of beating. If the beating is slow it is not unpleasant; witness, for example, the effect of the organ stop known as the voix celeste. This stop contains a range of pipes tuned slightly sharper than the pipes of an associated stop with which it is used. The result is to impart a wavy effect to the sound heard. As already stated, for notes in the treble clef about a semitone apart, difference of frequency produces beats whose rate causes the maximum unpleasantness in the sensation of beating felt by the ear. The sensation is less unpleasant if

[1] The structure of the ear is described in text-books on sound. It is there stated that the *cochlea* contains 3,000 rods of Corti, as well as the *basilar membrane* consisting of from 18,000 to 24,000 fibres, radially stretched and varying in length. A nerve filament from the brain is supposed to be connected to each fibre. On the piano there are only 88 notes.

the notes are a tone apart, and it disappears for notes which are about a minor third apart.

We may calculate the rates of beating which give these effects. The frequency of c″, in the treble clef, is 512. A note a semitone below this will have a frequency of $512 \times \frac{15}{16}$, or 480. The difference of frequency is 32, and consequently the vibrations will get completely out of step and completely in step again 32 times in a second. The ear will hear beating with the note c″ at this rate. The note a major tone below c″ will have a frequency of $512 \times \frac{8}{9}$, or 456; and the rate of beating with c″ will be 56. This is less unpleasant than beating at a rate of 32. The note a minor third below c″ will have a frequency of $512 \times \frac{5}{6}$ or about 427; and physical beats will be produced at a rate of 85 a second. This is about the limit of the sensation of beating. No unpleasant feeling is caused. The condition is comparable to the disappearance of flicker in a modern cinema.

These frequencies, and the intervals which give the most unpleasant sensations of beating, alter as we move to higher or lower pitches. If the student calculates the difference of frequencies for the following intervals in just intonation he will find that they are all 32.

FIG. 13

Yet as we move down the series we reach intervals which sound successively less and less rough, and the final one is a consonance. We are, of course, considering only pure tones at the moment.

On the other hand, when we are dealing with pure tones—such as the notes of a wide stopped pipe on the organ, which are practically pure tones, as we shall learn in Chapter IX—the semitone is not the roughest interval between notes low in the bass. At a pitch near the bottom of the piano, or that of the bottom notes of a 16-ft. stop on the organ, a major third produces a greater sensation of roughness than a semitone. On the piano the effect is partly obscured by the upper partial tones, the octave harmonic being powerful: it is more easily observed with a soft bourdon on the organ. Similarly, with notes considerably above the treble clef a quarter-tone would be a rougher interval than a semitone. As Helmholtz observes: 'The roughness arising from sounding two tones together depends, then, in a compound manner on the magnitude of the interval and the number of beats produced in a second.'

In the previous chapter we calculated the rate of beating produced by the difference tones of mistuned fifths and octaves when the notes of these intervals were pure tones. Helmholtz's conclusions call attention also to beating which occurs between the upper partial tones of two notes which just fail to be concords because they are separated by a mistuned interval. They also throw light on the acoustical qualities of what are known in academic counterpoint as imperfect concords or as discords. Helmholtz stated his conclusions thus:

'When two musical tones are sounded at the same time, their united sound is generally disturbed by the beats of the upper partials, so that a greater or less part of the whole mass of sound is broken into pulses of tone, and the joint effect is rough. This relation is called *Dissonance*.

'But there are certain determinate ratios between pitch numbers, for which this rule suffers an exception, and either no beats at all are formed, or at least only such as have so little intensity that they produce no unpleasant disturbance of the united sound. These exceptional cases are called *Consonances*.'

It is important to observe that these conclusions refer only to the physiological aspect of notes heard by the ear, that is, to the sensations they cause, disregarding any psychological effect these may have. What the art of music has to say about them we shall see later in this chapter; but notes which form good consonances will always sound 'in tune'.

Let us now examine in the light of these statements the effect, due to upper partials, of mistuning in the case of those intervals which are classified in academic counterpoint as perfect or imperfect concords; and we will include the twelfth and also the fourth. As in the case of difference tones, we will follow Helmholtz and show the fundamental tones by minims and the upper partials by crotchets. We will show the upper partials for each note of the interval only so far up the harmonic series as is necessary to find a unison with a partial tone (that is, including the fundamental tone) of the other note. The result is given in

Fig. 14 below; and the unisons which will cause beating if there is mistuning are shown by the heavy horizontal lines.

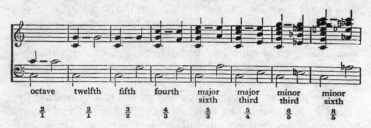

octave	twelfth	fifth	fourth	major sixth	major third	minor third	minor sixth
$\frac{2}{1}$	$\frac{3}{1}$	$\frac{3}{2}$	$\frac{4}{3}$	$\frac{5}{3}$	$\frac{5}{4}$	$\frac{6}{5}$	$\frac{8}{5}$

Observe that the unisons in the interval of a major sixth, ratio $\frac{5}{3}$, occur between the *fifth* partial of the lower note and the *third* partial of the upper note of the interval; and so on.

Fig. 14

We may set out in tabular form the rate of beating heard if we raise the pitch of the bottom note by 1 vibration a second and calculate the frequencies of the top tones in each bar of the above diagram, and the rate of beating between them. The result is as follows:

Interval	Beating frequencies		Rate of beating
Octave . . .	258	256	2
Twelfth . . .	387	384	3
Fifth . . .	387	384	3
Fourth . . .	516	512	4
Major Sixth . . .	645	640	5
Major Third . .	645	640	5
Minor Third . .	774	768	6
Minor Sixth . .	1,032	1,024	8

The number of beats is the same as the number of the partial tone of the lower note which ought to be in unison with a partial tone of the upper note. On the other hand, if we raise the pitch of the upper note by 1 vibration a second we obtain a rather different result.

Interval	Beating frequencies		Rate of beating
Octave . . .	256	257	1
Twelfth . . .	384	385	1
Fifth . . .	384	386	2
Fourth . . .	512	515	3
Major Sixth . . .	640	643	3
Major Third . .	640	644	4
Minor Third . .	768	773	5
Minor Sixth . .	1,024	1,029	5

The number of beats is the same as the number of the partial tone of the upper note which ought to be in unison with a partial tone of the lower note.

From these two tables we can deduce a simple rule for calculating the effect of mistuning, by any amount, any interval whose ratio we know. Thus from the second table, for a fifth, whose ratio is 3 : 2, *if we mistune the upper note by one beat a second the rate of beating produced is the lower number of the ratio*, that is, 2 per second. For the twelfth, ratio 3 : 1, it is 1; for the major third, ratio 5 : 4, it is 4. The rate of beating is always the lower of the two numbers in the ratio of the interval; and the mistuning may be either up or down without affecting our arithmetical result. If however, as in the first table, *we mistune the lower note by one beat a second, the rate of beating is given by the higher number of the ratio*. Thus the effect, on the fourth, of mistuning the lower note is seen from the table to be 4, and this is the higher number in the ratio of the fourth 4 : 3. Similarly for the minor third, ratio 6 : 5, mistuning the lower note, up or down, by 1 beat a second will produce beating at the rate of 6 beats a second. If our mistuning had altered the rate of vibration of the mistuned note by 2 vibrations a second all these numbers would have been doubled, and so on. For example, if we alter a major sixth, ratio 5 : 3, by mistuning its upper note to the extent of 6 vibrations a second the beating heard will be at a rate of 18 beats a second; or if we alter a perfect fifth, ratio 3 : 2, by mistuning the lower note to the extent of 3 vibrations a second, the beating heard will be at the rate of 9 beats a second.

We are particularly interested in the effect of the mistuning involved in the use of temperaments. In Chapter IV we calculated the alteration in the frequency of G in the treble clef caused by the mistunings produced in the mean-tone and the equal temperaments. The frequency of this note, in just intonation with middle C as 256, is 384. We found that when it was mistuned by one eleventh of a comma, as in equal temperament, its frequency became 383·57. When it was mistuned by a quarter of a comma, as in mean-tone temperament, its frequency became 382·8. The difference between those frequencies and that of just intonation are 0·43 and 1·2 respectively. Thus, applying our rule, the beating due to upper partials which is caused in the fifth by the mistuning required for equal temperament is at the rate of 0·86 or, say, 0·9 beats a second. For the mean-tone temperament it is at the rate of 2·4 beats a second. These are exactly the rates of beating found for the difference tones in Chapter IV. Neither of them is rapid enough to break up the volume of sound to an unpleasant degree though each is reinforced to a perceptible extent by beating between the difference tones if the notes are loud. If the music is moving at all rapidly the ear would hardly have time to hear any of these beats. We may compare this with the case of the major third. The frequency of e′ in equal temperament may be calculated by the method used for g′ in Chapter IV. Its logarithm will be found to be log 320+0·00343 (see p. 29 for this decimal), or 2·50858. From mathematical tables (see Table III, Appendix I) we find that this is log 322·54, which gives us 322·54 for the frequency of e′. By our rule, beating will therefore be at a rate of 2·54×4 or 10·2. This is becoming unpleasantly rapid, but as it is between the fifth and the fourth partial tones, respectively, of the two notes forming the interval, the beating will be rather faint. Further, it is not reinforced by any perceptible beating between the difference tones, for we saw in Chapter IV that it was only at the octave, and to a less degree at the fifth, that really noticeable beating due to difference tones took place between two pure tones.

The rate of beating due to the mistuning of the fifth in the equal and mean-tone temperaments has been worked out, not only to enable the student to picture more clearly the effect of temperaments, but because it affords a means of tuning the organ to meet exactly the requirements of these temperaments. The student will be interested to find that he can now calculate the theoretically correct rate of beating for himself for tempered fifths of any pitch. To do so will help him to understand the instructions which manuals on tuning the organ give for 'laying the bearings' of an octave on the 4-foot principal. For example, one of the tuner's rules is that the beats produced by the tempered fifth above a given note are at the same rate as those produced by a tempered fourth below the given note. The reason for this appears from the rules we have found for calculating the effect of mistuning an interval. Suppose that our given note were c′. The rate of beating in the tempered fifth produced with g′ will be twice the error of frequency introduced into g′ by the temperament (which may be either mean-tone or equal temperament); for we must apply the rule for mistuning the upper note of our interval, ratio 3 : 2. The rate of beating in the tempered fourth which is produced by sounding g, an octave lower, with c′ will be four times the error of frequency introduced into g by the temperament; for we must apply the rule for mistuning the

lower note of our interval, ratio 4 : 3. But the error in the frequency of g′ is twice that of g, because the frequency of g′ is twice
that of g. Consequently the two rates of beating must be the
same if the octave gg′ is true.

We can apply this arithmetically by working out some beats
which would occur in tuning the organ in equal temperament.
The basis of correct organ tuning is the correct tuning of an
octave of the 4-foot principal on the Great Organ. If middle C
on an 8-foot stop had a frequency of 256, it would be the tuner's
aim so to tune the principal that when the notes of the octave
above middle C were played on the keyboard he would obtain
frequencies twice those given in the right-hand column of
Table IV, Appendix I. The rate of beating for the interval C to
G at this pitch is about 1·7 beats a second, for it is an octave
higher than the fifth whose rate of beating we calculated on
p. 53. G would then have a frequency of 767·13. If the student
now calculates the rate of beating for a tempered fourth taken
downwards from this G to a D with frequency 574·70 he will
obtain as a result a rate of about 2·6 a second. Actually it would
not be possible to tune an organ by such calculations alone, if
only because a small difference of pitch would throw them all
out. The fine tuning of an organ is a skilled craft, but the tuner,
in fact, uses a series of checks based on the same principles as
our calculations. His C would normally have a frequency rather
greater than 256. When his fifth C to G on the principal was
made 'narrow' he would produce rather less than two beats a
second. When the fourth from G down to D was made 'wide'
he would produce rather less than three beats a second. This
agrees with the tuner's rule; for if C to G gave two beats a
second, the fifth from G to the D above it would give about three
beats a second, since it is higher by a tempered fifth, ratio nearly
3 : 2. Proceeding thus, the tuner could find A by a fifth up from
D, E by a fourth down from A, and so on, checking the result by
a reverse process. When tuning the organ in mean-tone temperament the tuner used to produce about five beats a second in
the fifth from middle C to G on the principal.

We have been considering the effect of beating in intervals
which differ very slightly from consonances. The beats are
slow, and for the octave and the fifth they are readily perceived
since they are between the prime and the first harmonic overtone in the one case and the first and second harmonic overtones
in the other. The higher we proceed up the series of partial
tones the weaker they normally become. Consequently the

beating due to mistuning of thirds and sixths, though more rapid, and therefore reaching sooner an unpleasant rate as the mistuning increases, is of slighter intensity. Intervals such as the octave and the fifth, for which mistuning produces beating between the lower partial tones, are said to be *sharply defined*. The larger the numbers which form the ratio of an interval, the higher will be the partial tones which beat as imperfect unisons when the interval is mistuned, and the less sharp the definition of the interval.

Certain instruments produce notes which are lacking in harmonic overtones and consist of nearly pure tones, slightly reinforced perhaps by the octave harmonic. Such a note is sounded by a stopped organ pipe or a flute not blown loudly. A note of this nature may be said to lack definition in the sense that if used in accompaniments to vocal music it does not assert itself aggressively against loss of pitch by the singers. This is well known to organists. In soft accompaniment a pipe with an incisive note containing harmonics, as we shall see in Chapter IX, is a better support to the choir than a wide stopped pipe of 8-foot tone. To maintain the singing of a congregation up to pitch the principal and the fifteenth, which reinforce the upper partials of the notes played on the organ, are more important than the diapason tone itself.

We have seen that the octave is hedged in by dissonances which give it the sharpest possible definition. At the other extreme we have discords, like the tritone and the imperfect fifth, which contain so much dissonance due to beating between partials, even when they are tuned exactly,[1] that beating due to slight mistuning makes no perceptible difference. In Chapter III we saw that the ratio of the tritone was 45 : 32 (see p. 34). This means that we have to reach the 45th partial of the lower note and the 32nd partial of the upper one before we meet an exact unison to mistune. Obviously these partials are negligible, and if we proceed to lower partials we shall find that before we reach audible ones we have departed too far from anything like a unison for mistuning to alter the dissonance. That is why in Chapter III we referred to dissonant intervals as not having the same definition, due to beats between their upper partial tones when mistuned, as consonant intervals.

[1] The interval of a tritone could be tuned exactly by ear only as a happy accident. For reasons given above, the ear would have no means of estimating exact tuning. The ear would succeed better if it referred both notes to the dominant in turn. The tritone could be tuned exactly in the laboratory by electrical means.

In any case, if the mistuning of an interval is greater than we have been supposing in the preceding paragraphs, the beating due to imperfect unisons between partial tones becomes more and more concealed from the ear by the mass of sound, and by other dissonances of the primes which now become important. Let us consider these other dissonances, and for the purpose let us consider partials extending to the sixth, which is the highest partial tone of any significance in the notes of the piano. Let us then set out in staff notation, as in Fig. 15, side by side, the two primes represented by minims, with their partial tones above them represented by crotchets; and let us mark with a dotted line partials which are a tone apart and with a continuous thicker one partials which are a semitone apart. The result will be as follows:

FIG. 15

In estimating the degree of roughness of the beating partials, allowance should be made for their rapidly diminishing intensity as we ascend the harmonic series; and, as between the fourth and the major sixth, it should be remembered that the interval g'f' is a major tone, that between a'g' a minor tone; though if the fifth partial of the lower note were powerful the fourth would lose its acoustical advantage over the major sixth. The octave and the twelfth are perfect consonances—as must be any interval between a note and one of its own harmonic series. The major tenth is a perfect consonance within the limits of the partials selected. It would not be so perfect as the twelfth if we included two more partial tones of the lower note, both within beating distance of a partial tone of the other. The fourth, on the other hand, is a better consonance than the eleventh, for the upper prime of the eleventh is itself within beating distance of the third partial tone of the lower prime, while its second partial tone beats strongly with the fifth partial tone of the lower prime.

The improvement in the consonance of a major third which results from raising by an octave the upper note of the interval, and the contrary effect with the fourth, are interesting. The general observation with which we began this chapter would lead us to classify intervals for consonance by the size of the numbers which give their ratio. This classification, which is, of course, an acoustical one—not that of the art of music—would lead us to arrange intervals of an octave or less in the following order 2 : 1, 3 : 2, 4 : 3, 5 : 3, 5 : 4, 6 : 5, 8 : 5. If an interval, such as a third, has a ratio in which the smaller number is even, we divide that number by raising the upper note an octave. Thus a tenth has the ratio 5 : 2, and should be placed earlier in our series of ratios than the third, 5 : 4. On the other hand, if an interval, such as a fourth, has a ratio in which the smaller number is odd, we multiply the larger number by raising the upper note an octave. Thus an eleventh has a ratio 8 : 3, and must be placed later in our series of ratios than the fourth, 4 : 3. That the general observation leading to our series is sound, judged by Helmholtz's physiological test for consonance, is confirmed by examination of all the other intervals in Fig. 15.

It is evident that, from an acoustical point of view, the number and intensity of the upper partial tones in a note are very important in considering the dissonance of any interval formed with another note. In Chapter IV we realized the necessity of understanding the nature of the vibrations which cause sound. Here is another example of the necessity. What are the conditions which effect the number and strength of the partial tones in a note? This question we will examine in the next chapter, and in Chapters VIII, IX, and X.

Meanwhile it is interesting to observe that if a musical instrument is so designed that its notes contain upper partial tones extending over a considerable range and sounding at all strongly, we should expect each note of the instrument to contain its own element of dissonance through beating between the higher partials. For example, the 15th and 16th partials are a semitone apart. The notes of such instruments will sound rough or harsh.

One other observation should be added. The effect of pitch on the dissonance of an interval is important. As we proceed up the scale from the bottom, the interval which causes the maximum irritation to the nerves of the ear through the sensation of beating becomes smaller and smaller. But the decrease, as we have seen, is not at all proportional to the increase of frequency: it is much less. This explains why a major third sounds so much less consonant in the bass than in the treble clef. As shown in Fig. 15, a major third between e and c in the bass produces beating between two partials a semitone apart at the pitch at which beating at this interval attains its maximum unpleasantness. If the pitch of the primes were raised two octaves, the interval which would produce maximum unpleasantness at the pitch of these partials would be considerably less than a semitone. The rapidity of the beats between the offending partials would be nearer the stage of inoffensiveness. For similar reasons the fifth shown in Fig. 15 would be a better consonance if it were raised an octave.

We have still to examine the dissonance of triads obtained by combining two intervals, when we shall find that Helmholtz's physiological criterion applies precisely the same test as is applied in practice by the ear of a violinist to make him play 'in tune' in a string quartet. We arranged consonant intervals within the octave in the following order arithmetically; and we saw from Fig. 15 that the order accorded with that obtained by Helmholtz's physiological test: $\frac{3}{2}, \frac{4}{3}, \frac{5}{3}, \frac{5}{4}, \frac{6}{5}, \frac{8}{5}$. We have shown them as fractions for ease in arithmetical manipulation. If we take these intervals in pairs and multiply each pair, we shall find, by trial, that only three pairs give, as their product, a fraction which is in our series; namely, $\frac{5}{4} \times \frac{6}{5} = \frac{3}{2}$, $\frac{4}{3} \times \frac{5}{4} = \frac{5}{3}$, and $\frac{4}{3} \times \frac{6}{5} = \frac{8}{5}$, and we may take the fractions to be multiplied in the order shown or in reverse order and obtain the same result. We thus obtain six triads which are acoustically satisfactory; and we may show them, with the same note for bass, in staff notation as in Fig. 16 overleaf.

FIG. 16

We will examine these triads in turn for beating between upper partials and also for their first difference tones.

Beating between the upper partials is shown by the following diagram:

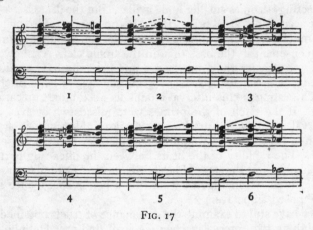

FIG. 17

The first difference tones are shown by the following diagram, in which, for convenience of reference, we will represent the triads as belonging to the keys of C major and C minor.

FIG. 18

In these diagrams 1, 2, and 3 are the major triad and its so-called inversions; 4, 5, and 6 are the minor triad and its inversions. Taking first the major triads of Fig. 17, it is evident that 2 is considerably less dissonant than 3. At a higher pitch, which will diminish the sensation of beating of intervals of a tone between the partials, 2 will be smoother acoustically than 1; 3 is surprisingly dissonant. The effect of these triads is in no way prejudiced by the first difference tones, which are all natural to the harmony. Taking now the minor triads, they all exhibit dissonance, though 5 is rather less dissonant than 3, while 6 is clearly the most dissonant chord of the whole six. More-over, the harmonic sense of 4, 5, and 6 is interfered with by the first difference tones, which introduce harmony alien to that of the triads. This gives to the minor triads the veiled effect, to use Helmholtz's term, which is characteristic of minor harmony.

How does the dissonance of these triads bear on the use of temperaments? We have seen that mistuning of the fifth has a more serious effect, due to beating between the partials and between the first and second difference tones, than mistuning of the third. The fifth is nearly correct in equal temperament, but the major third is $\frac{7}{11}$ of a comma sharp. It is sometimes stated, as an inference from these facts, that equal temperament tampers most with those intervals which can best stand it. This is undoubtedly true of melody, or of two-part writing, but where there are more than two parts Helmholtz will have none of it. The statement, if not qualified, would prove too much. It would show that equal temperament was better for organs, in all suitable keys, than the mean-tone temperament; and this is clearly not true or the mean-tone temperament would not have kept its vogue. Consider the diagram for the first difference tones in 1 of Fig. 18. The major third and the minor third in this triad both produce the same first difference tone. Now we saw in Chapter IV that if an interval is increased the first difference tone rises in pitch; if it is diminished the first differ-ence tone drops in pitch. In equal temperament, the third in the major triad being too sharp, the first difference tone of the major third will be sharpened, but the first difference tone of the minor third will be flattened. These first difference tones will beat per-ceptibly. In fact, their rate of beating will be 5·5. As Helmholtz observes: 'In a consonant triad every tone is equally sensitive to false intonation, as theory and experience alike testify, and the bad effect of the tempered triad depends especially on the imperfect thirds.'

As a method of tuning keyed instruments, equal temperament has established its position for reasons other than this. Its great advantage is the unlimited facility of modulation it affords to the keyed instruments; and with that facility must be combined the varying degree in which the keyed instruments of the orchestra command some control over intonation, and to that extent enable the player to express his feeling for playing 'in tune'. Observe the bearing on this of Stanford's insistence on the study of the pure scale referred to in the Introduction. The value of this study for orchestral composition is explained in the work quoted. The further quotation from Parry at the end of Chapter II illustrates the importance of the contribution made to the modern scale system by tuning in equal temperament.

The full significance of Helmholtz's physiological investigation of consonance, dissonance, definition, and the effect of mistuning is now evident. Like Stanford, Helmholtz observes that good orchestras do not play in equal temperament.[1] So far as the keyed instruments are concerned, the performance of an orchestral composition may be conceived as having a skeleton of equal temperament. But, in the flesh and blood with which the instruments of the orchestra clothe it in a manner dependent on the adaptability of their notes, the body of music comes to life in just intonation with an approximation that is measured by Helmholtz's criterion of consonance: the player's musical instinct is to play the 'essential notes' in tune, and the less the dissonance the more important to the musician's ear this instinct becomes. Moreover, if an instrument adjusts a note by a comma, that comma may easily disappear, without upsetting the ear, in a discord which is lacking in definition owing to its dissonance. As the physicist would say, the power of the ear to measure intervals depends on its own limits of accuracy, and those limits depend, in turn, on the duration of the note, the definition of the intervals, and the dissonances of the harmony within which the ear has to do its work. To this must be added one other factor: as we have seen, the smoothness of a consonance varies with the pitch. It is somewhat less in the bass than in the treble. We see now why we do well to think of a scale *system*, and why it is as unsound scientifically as it is wrong musically to conceive of scales as if they fixed the position of unessential notes or the ratios of dissonant intervals with extreme arithmetical nicety.[2]

[1] See pp. 37 and 125.   [2] See also the note in Appendix I, p. 149.

In passing, it should be noted that the ear of a musician who is constantly playing an instrument with a rigidly fixed intonation, such as the organ, not infrequently loses some of its sensitiveness for consonance, that is for absence of beating, because it is so often listening to music which is not exactly in tune.

The art of music has a method for its procedures based on experience. To acquire a good technique for writing music is to gain a mastery of that method which becomes instinctive. At the other extreme that method distinguishes singing from an unmusical sound of changing frequency like the howling of a dog. Fundamentally, it rests on the selection of intervals which the ear deems appropriate for varying melodic or harmonic requirements. It is not confined to filling the octave with a number of notes of rigidly fixed intonation: that they must be so confined is the inescapable defect of the keyboard instruments. A passage of Palestrina or Byrd played on a harmonium, even one tuned to the pure scale, can never reproduce the notes to which it is sung by a good choir. Mean-tone tuning was adopted, in the sixteenth century, as a compromise because the notes of virginals and organs were fixed, not mutable. The eventual result of the extending use of modulation in the seventeenth and eighteenth centuries was to replace mean-tone tuning by equal temperament, as a better approximation to new requirements. But the modern scale *system* is a flexible thing, as was the pure scale of the sixteenth century. Good orchestral players make corresponding use of such flexibility of intonation as their instruments command in actual performance.

With instruments whose notes are rigidly fixed and tuned in equal temperament the conditions differ from those of the brass or of the woodwind instruments of the orchestra. In the pianoforte, as we saw in Chapter IV, the difference tones are only heard momentarily since the sounds of the instrument are loud only at the moment when the hammer strikes the strings. Further, the partial tones above the sixth are practically eliminated in modern pianofortes by the design of the hammer, the selection of the point at which it strikes the string and the duration of the impact—which also have the effect of weakening the partials above the second or third. To this it should be added that the notes of pianoforte music usually succeed each other rapidly, as the instrument is deficient in sustaining power: they then leave the ear too little time to hear comparatively slow beating.

The instrument which suffers most is the organ. The difference tones of the thirds are audible, save with the softest stops, and the reed stops contain higher partials than appear in the pianoforte. The addition of mixtures aggravates the trouble. If the mixtures were not tuned in just intonation for any given note, they would produce dissonance for that note by itself. Consequently they must inevitably contribute to the dissonance of chords played in equal temperament by supplying the higher partial tones which emphasize the dissonance. As Helmholtz observes: 'Every chord furnishes at once tempered and just fifths and thirds, and the result is a restless blurred confusion of sounds.' To this observation organists may well retort that it leaves out of account one of the most important stops of the organ, a stop not to be found in the builder's printed specification. That stop is the acoustical quality of the building, the effect of which is partly psychological. Sir Walter Parratt is said to have remarked that even a sneeze would sound musical in St. Paul's Cathedral.

The examination of the dissonance of the major triad and its inversions in just intonation assigns to these chords a different place from that assigned to them by the art of music. It shows that the major chord of the 6/4 is acoustically the smoothest consonance, yet for centuries the art of music has treated this chord as a discord.[1] There is a significant sentence in Helmholtz's work: 'The dispute as to the consonance or dissonance of the fourth has been continued to the present day.' This dispute is, in effect, whether music is an art or a science. No room for controversy would have been left had the test of the disputants been that of the use of the 6/4 by Bach in his forty-eight preludes and fugues for the clavier. Doubtless it is the acoustical qualities of the 6/4, supported perhaps by the practice of regarding it as an inversion of a concordant triad, which explain the need for a sentence in Buck's *Unfigured Harmony*: 'An early opportunity must be taken of warning the student against the use of the 6/4.' He continues: 'Instances might be quoted where it is beautifully used as . . . a point of colour.' The instances he has in mind, as illustrated by a reference to a Prelude of Scriabin's, are those in which composers use it without regard for the 'rules' for its resolution as an 'essential discord' or its treatment as unessential harmony.

The concords of three-part counterpoint are:

(*a*) bass note+major or minor third+perfect fifth, from the *bass*.

[1] This chord is marked 2 in Figs. 17 and 18.

(b) bass note+major third+major sixth, from the *bass*; bass
note+minor third+major sixth, from the *bass*; bass note
+minor third+minor sixth, from the *bass*;

and the upper parts may be arranged in either order or at any
distance. Not only does the art of music, as derived from the polyphonic period, therefore exclude the chord of the 6/4 from the
available concords, but it includes a chord of the 6/3, DFB in
the key of C major, which since the beginning of the polyphonic
period has been accepted as a concord, but which is not included
in the list of acoustically consonant triads. The reason is not far
to seek: in this chord not only is the interval FB a tritone but,
as we saw in Chapter I, the interval DF is less by a comma than
a true minor third. While D, in practice, is a mutable note the
'definition' of the major sixth DB, in contrast with the dissonances of the triad, suggests that the musical ear will wish to
hear this interval true.

Observing that the fourth, fifth, and sixth partials and their
octaves formed the notes of the various inversions of the
common chord, relationships exhibited in effect by 1, 2, and 3 of
Fig. 18, Rameau (1683–1764) and after him D'Alembert (1717–
83) were led to formulate theories of harmony which had a
physical basis. These theories resulted in the conception of a
fundamental bass, the prime of these partials, which in the hands
of their successors developed into a system of roots. Their contributions to the theory of harmony were incomplete, because
their material was scanty and the physiological data which
Helmholtz contributed were lacking. Yet the value of their contributions should not be under-estimated, for their attempt had
the advantages of shifting the problem from the metaphysical
ground, accepted as we have seen by Euler, to a basis of physics,
and of replacing vague, confused theory by orderly system. That
uncritical development of their conception at other hands
should have had unfortunate results does not detract from the
merit of their achievement, to which Helmholtz pays generous
tribute. There was, indeed, little excuse for the extent to which
these developments ran. The relationships of the minor triad,
for instance, suggest no fundamental bass. No hypothesis can
be sound which proves too much. Rameau's invention of
'generateurs' led to the search for roots for every conceivable
chord. The augmented sixth was granted the privilege of two
roots. At the hands of theorists in this country, the chord of
the added sixth, which occurs so naturally in Palestrina—as at
the end of the Christe Eleison in the Missa Papae Marcelli—

was catalogued as a dominant eleventh with the root and third missing. As Dr. Pole pointed out in a series of lectures on Helmholtz's theories delivered at the Royal Institution in 1877, it seemed unnecessary to stop at the elimination of the root and the third: why not eliminate the fifth also, and so exhibit the major triad on the subdominant as a dominant discord?

The matter is set in its right perspective to-day. Thus Tovey, writing of the practice of playing from figured bass, says:

'It is an empiric craft. But it had the misfortune to become a science, when, early in the 18th century, Rameau discovered the theory of the fundamental bass. This is an imaginary bass (best when most imaginary) that gives "roots" to all the essential chords of the music above it. The conception is true only of the most obvious harmonic facts; beyond them it is as vain as the attempt to ascertain your neighbour's dinner from a spectrograph of the smoke from his chimney. . . . In England Rameau's doctrine raged unchecked by taste or common sense, and culminated in Dr. Day's[1] famous application of homoeopathy to the art of music . . . The remedy lies in cultivating vivid impressions of the actual relations between counterpoint and harmony in detail, between tonality and form in general, and between key-relations and chromatic chords.'[2]

To this quotation Helmholtz's criterion of consonance supplies the acoustical complement. Concerning itself only with the sensations in the ear of the musician, it brushes aside all conjecture which leaves the ear out of account, which ignores the evidence of the history of musical composition, or which constructs, from the notes of the harmonic series, what Donkin described[3] as 'the theory of artificial scales', with fixed notes. For such conjecture Helmholtz's criterion substitutes the simple question: How does the ear tell whether two notes are in tune? So far as it lies in the power of acoustics to do so, the answer explains the elements of the scale systems known to the art of music for the last four hundred years or more, on which the importance of counterpoint in musical studies must ultimately depend.

[1] Alfred Day, M.D., 1810–48, practised in London as a homoeopathist.
[2] *Encyclopædia Britannica*, 14th ed., article on 'Harmony'.
[3] *Acoustics*, W. F. Donkin, p. 22.

VI

THE NATURE OF SOUND, THE QUALITY OF MUSICAL NOTES

A PICTURE of the way in which sound is transmitted from its source to the ear by the air can be formed if one considers what happens when the prong of a tuning-fork is vibrating. When the two prongs are moving away from one another each prong presses against the air which is in its way and compresses it. The compression of this air communicates itself to the surrounding air, and consequently a pulse of compression travels out in ever widening circles. When the prongs move towards each other the opposite motion takes place. The air which was previously compressed now has to fill the space left by the retreating prong and this causes a rarefaction. This, in turn, travels out into the surrounding air in ever widening circles. In this way a regular succession of condensations and rarefactions occurs in the air along a line from the tuning-fork to the ear. When it reaches the ear it travels down it to the drum, which it causes to vibrate in and out in accordance with the motion of the air. This produces the sensation of a musical note in the nervous system of the ear.

To prevent misconception, it should be pointed out that this description of the sound wave produced by one surface of a tuning-fork, though correct as far as it goes, is incomplete as a description of the sound wave produced by the whole fork. One has to hold a tuning-fork close to the ear to hear its note. To make it sound a loud note it is necessary to insert its stem in a resonance box or to hold it down on a table or some other piece of furniture. This will be explained in Chapter VIII; while the combined effect of the four vibrating surfaces of a tuning-fork held in the hand, the outside and inside of each prong, is described in Chapter VII.

Waves of sound therefore consist of alternate condensations and rarefactions of the air; the particles of the air may be conceived as oscillating backwards and forwards in the line along which the sound is travelling from the source to the ear; and similar oscillations take place along all other lines radiating from the source, though these, of course, do not reach the same ear.

The way in which the particles of air move if the sound is a pure tone is both interesting and important. It is possible to calculate by mathematics the exact way in which each particle of air must move; and the motion is found to be the same as that of a steadily moving pendulum swinging through a very short distance, so short that its motion is indistinguishable from motion along a straight line. For this reason Helmholtz refers to the vibration, caused in the air by a pure tone, as a pendular vibration. Another name for the motion is *harmonic motion*; and it is the simplest form of vibration, regarded mathematically, that a particle can have.

A mental picture of harmonic motion can be made if one thinks of a weight, at the end of a string, whirled some distance away in a horizontal circle at the level of one's eye. To the eye the weight will appear to move backwards and forwards along a horizontal line. It will appear to move very rapidly in the middle of its swing, but increasingly slowly as it reaches the end of its swing, when it stops and turns back with ever increasing speed till it comes to the middle position again. It repeats the process on the other side of the middle position. An example of this motion can be seen in a coupling-rod on a railway engine at some little distance when it is running into a station on the platform of which the observer is standing. The coupling-rod appears to move up and down in a vertical line, and its movement is very rapid in the middle and slow at the top and the bottom.

The motion can be represented graphically if a plan is made of the movement of the weight which was supposed to be whirled in a horizontal circle. Such a plan is drawn in Fig. 19; the eye being supposed to be in the plane of the paper and at a considerable distance beyond the bottom of the page.

Let it be supposed that our weight, represented by the letter P, is whirling round this circle in a counter-clockwise direction shown by the arrow. The eye will see the circle edgeways as a straight line CD; and the weight will appear to move backwards and forwards along this line, rapidly at the middle point O, then increasingly slowly as it approaches C, where it stops, reverses its motion, and travels with increasing speed back to O. The same motion will then be repeated to D and back.

When the weight is in any intermediate position such as P, it will appear to the eye to be at N, where PN is perpendicular to CO. Its distance away from O will appear to be equal to PM, for the eye is supposed to be so far away that NO and PM appear the same length. Now, mathematicians have calculated the length of PM to many places of decimals for a great many sizes of the angle *POA*. If they assumed the length of *PO* to be 1, they would say that *PM* was equal to what they call the *sine* of the angle *POA* and they have made tables of these sines. We can therefore use these tables to calculate the displacement of N at any fraction of the time which *P*, moving with uniform speed, takes to go round the circle. A few of these sines, calculated to five places of decimals, are set out in Appendix I; and we can use them to calculate the lengths of *ON* for a number of positions of *N*, and to show the result as decimals of *OC* or *OD* which will each have the value 1 on this basis of calculation.

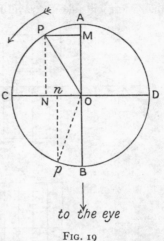

FIG. 19

The student who has an elementary acquaintance with science and has done the binomial theorem in algebra will find it interesting to turn, at this point, to Appendix III, where he will find an explanation why the motion of the air for *very small displacements* due to a pure tone is what Helmholtz called a pendular vibration or what is often called harmonic motion.

Now suppose that we have a number of particles of air lying on a line along which the sound of a pure tone is travelling. We will represent these diagrammatically; but to keep the diagram from becoming too detailed we will consider only particles in the nine positions which *n* would occupy in Fig. 19 when *pOB* was in turn 0°, 22½°, 45°, 67½°, 90°, 112½°, 135°, 157½°, and 180°. These are the positions read backwards through which *n* would pass in its excursion from *O* to *C* and back. We shall see in a moment why they are taken backwards. There would be eight similar positions to record for the excursion of *n* to *D* and back; and the last one, the seventeenth in all, would be the position from which *n* started in its excursion to *C*.

The first nine particles are represented in Fig. 20. In (i) they are all at rest as they would be when not disturbed by any sound. In (ii) they are supposed to be disturbed by the vibrations of a pure tone moving from left to right as shown by the winged arrow. Each particle will then start swinging forwards and backwards as it is disturbed alternately by condensations and rarefactions. The time taken by a condensation to travel from the position of 1 to that of 9 in (i) will be that which particle 1 will take to make half a complete swing, that is, the time p will take to travel half-way round the key circle. The time a condensation will take to travel from position 1 to position 2 in (i)

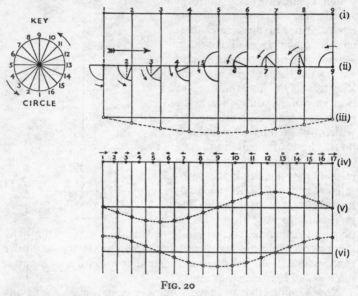

FIG. 20

will be $\frac{1}{16}$ of the time of a complete swing; the time it will take to

travel from position 1 to position 3 in (i) will be $\frac{1}{8}$ of the time

of a complete swing; and so on. Consequently, if at a given moment particle 1 is in the position corresponding to position 1 of p in the key circle, particle 2 will at that moment be in the position corresponding to position 2 of p in the key circle, and

so on. The motion of each particle will be $\frac{1}{16}$ of a phase, to use

the scientific term, ahead of that of the next particle shown on its right.

This enables us to construct the position of each particle in (ii) of Fig. 20; for which purpose we have drawn for each of them a quadrant of the key circle. The maximum distance which each particle swings to right or left of its equilibrium position is called the *amplitude* of the vibration. When we come to particle 5 we find a particle at the end of its swing, and therefore at a maximum distance to the left of its equilibrium position as shown in (i). When that happens we find particle 1 in the middle of its swing with no displacement, particle 2 displaced backwards by 0·38 of the amplitude of its vibration (for sine 22½° is 0·38 to two places of decimals) but moving to the right, particle 3 by 0·7 of its amplitude and also moving to the right, particle 4 by 0·92 of its amplitude and moving to the right, particle 6 by 0·92 of its amplitude but moving to the left, particle 7 by 0·7 of its amplitude and moving to the left, and so on.

If we continued the diagram in (ii) farther to the right we should find particles 10, 11, 12, &c. in the positions and with the motions, reversed, which images of 8, 7, 6, &c. would have in a mirror held in the position of particle 9. Thus particle 10 would be displaced to the right by 0·38 of the amplitude of its vibration and be moving to the left. Particle 11 would be displaced to the right by 0·7 of its amplitude and would be moving to the left, and so on. Finally, particle 15 would be displaced to the right by 0·7 of its amplitude but would be moving to the right, particle 16 would be displaced to the right by 0·38 of its amplitude and would be moving to the right, while particle 17 would be in its equilibrium position and be moving to the right. All the 17 particles are shown in (iv) of Fig. 20 and the arrows over them show how they are moving, the lengths being an approximate guide to the rate at which they are moving. A small nought over a particle means that it is standing still.

Now it is obvious that if we look at the positions of the particles in (iv) of Fig. 20 it is not easy to make an exact comparison of the extent to which the several particles are displaced. As a means of graphical representation (iv) would not be very convincing. Still less would it be convincing if the displacements were really drawn to scale. We can see that this is so if we consider how long the distance between particles 1 and 17 must be in practice. Let us suppose that the pure tone is middle C, with 256 vibrations a second. If the distance between particles 1 and 17, or the wave-length, to use the technical term, is λ, it follows that 256λ will be the distance the sound travels in a second. Now, sound travels about 1,100 feet a second at normal temperature, so that λ would be $\dfrac{1,100}{256}$ feet, or about 4 feet 3 inches. It is obvious that since the displacement of the air in the ear does not cause an uncomfortable pressure on the drum it must be trifling compared with this figure. In fact, the maximum displacement of the air in the ear would be of the order of 0·00001 inches for a note like middle C heard as of average loudness. This would make it impossible to represent graphically, by a series of particles like those in (iv), the actual displacements in a way which would make any appreciable impression on the eye. Accordingly a graphical device is adopted. In (iii) the displacements are shown by a convention. The displacement of particle 4, say, to the left is shown by an equal displacement downwards of particle 4 in (i). The displacement of particle 8, say, to the left is shown by an equal displacement downwards. Moreover, the displacements downwards of particles 2 to 8 are proportional to the sines of angles $22\frac{1}{2}°$, $45°$, $67\frac{1}{2}°$, $90°$, $112\frac{1}{2}°$, and so on. The curve drawn through the points which represent the displacements is called a *sine curve*. It can readily be reproduced by plotting a series of sines on squared paper. We will come back to this.

In (iv) the displacements of particles 10 to 16 to the right of their equilibrium position are shown. Our convention requires us to show these upwards as is done in (v). Observe that in (v) we have exaggerated the displacements by doubling them. Even then, though the displacements in (iv) are already greatly exaggerated to begin with, it is not easy to see the true character of the freehand curve in (v). Accordingly, it is convenient for the purpose of drawing this curve to exaggerate the displacements a very great deal by drawing them to a very exaggerated scale: just as sections of a map are drawn on a scale which shows elevations many times larger than corresponding distances on the level.

The curve which we can draw by substituting upward and downward displacements (on a very exaggerated scale) for actual displacements to the right and left is called an *associated displacement curve*. To draw associated displacement curves we may use the sines of the angles given in Appendix I. It will be observed that the sines of the angles $18°$, $27°$, $45°$, $54°$, $72°$, and $90°$ all happen to have 0 or 5 in the second place of decimals. This is very useful in drawing associated displacement curves on squared paper—which as supplied by stationers is ruled with 10 lines to the inch or centimetre—when we shall use a very exaggerated scale for the displacement. The curve is obtained by drawing a freehand line through the points which are circled in the diagram shown in Fig. 21.

The distances along the axis AB are multiples of 9°, the distances of the points in circles below or above AB, are the sines of the angles 18°, 27°, 45°, &c., taken from the table in Appendix I. To show, on the same scale, vibrations of greater or less amplitude, representing sounds which are louder or softer than that which Fig. 21 represents, it is necessary only to increase or diminish in the same ratio all the decimals taken from the sine table in Appendix I. It is easy, on squared paper, to double or halve or quarter the amplitude. Similarly, to show vibrations of a different frequency, and therefore a different wave-length, the

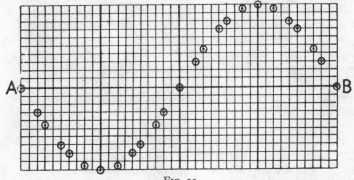

FIG. 21

distances along AB are altered proportionately. If the frequency is doubled, the wave-length is halved, and so forth.

The method of drawing an associated displacement curve has been worked out in this chapter in great detail, because it is very important that the student should be equipped to draw such curves for himself on squared paper, with full understanding, to represent the nature of the disturbance caused in the air not only by a pure tone, but—as will be explained later—by a complex note containing partial tones, and to be able to draw curves which represent the disturbance produced in the air by sound vibrations which are moving in opposite directions. The student is strongly urged to draw such curves for himself, as occasion arises, in order to obtain a thorough grip of the acoustical facts dealt with in the rest of this chapter, and in Chapter IX. In fact, once we had found the nature of the vibration caused by a pure tone, the principal aim of what has followed in this chapter so far has been to help the student thoroughly to understand what associated displacement curves mean, and how to draw them.

Before dealing with disturbances in the air such as those arising as described in the preceding paragraph, we may make a final observation on two other matters illustrated in Fig. 20. Obviously in (ii) and (iv) particle 1 is at a position of maximum condensation and particle 9 at a position of maximum rarefaction. Thus, remembering that the wave is supposed to be moving from left to right as shown by the winged arrow over (ii), we see that maximum condensation occurs when particles are moving forward with maximum velocity in the direction in which sound is travelling. Maximum rarefaction occurs when the particles are moving with maximum velocity in the opposite direction. As already stated, the velocity of these motions is indicated in both direction and amount by the direction and length of the arrows in (iv). As explained in Appendix IV, it is possible to represent the alterations in the condensation and velocity by associated curves, both similar to that which is shown at (vi) in Fig. 20.

We may now proceed to draw associated displacement curves to find out what happens to the particles of the air when they are disturbed at the same time by the sound of two or more pure tones travelling in the same direction. Let us begin by drawing the associated displacement curves for two pure tones a fifth apart. We will suppose, for convenience, that the amplitudes of the two vibrations are the same;[1] and we will also suppose that they are in step to begin with at a point of no displacement. If the vibrations caused by the bottom note are represented by a sine curve with a thin line, and those for the top note by a sine curve with a dotted line, the result will appear like (i) in Fig. 22, in which the sound is supposed to be moving from left to right as shown by the arrow.

Except when they show the same displacement at the same time, the two curves would represent each particle of air as being in two places at once, which is impossible. Huygens (1629–93) first enunciated the principle of superposition which tells us what happens to each particle. Suppose one vibration would have the effect of moving the particle through a distance x to the right, while a second one would have the effect of moving it through a distance y to the right. The total effect would be to move it $x+y$ to the right. But if the second vibration would have the effect of moving it y to the left the total effect would be to move the particle $x-y$ to the right; or, if y were

[1] By so doing we shall be representing the upper one as louder than the lower one (see Appendix V), but for present purposes this is not material.

greater than x, $y-x$ to the left. This enables us to fix the position of every particle, by algebraic addition of the displacements upwards or downwards of the two curves. We can easily perform these algebraic additions by counting the lines on our squared paper for two simultaneous displacements. A little practice will enable the student to do this quite quickly. He is

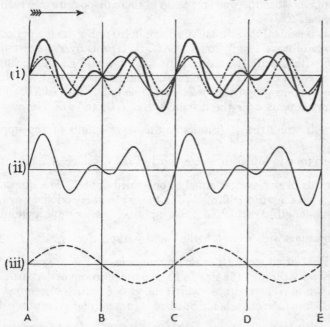

FIG. 22. Diagram of the sound wave, in (i) and (ii), produced by two pure tones a fifth apart.

recommended to do this himself, on squared paper, for the two curves under consideration, and to compare the result with (i) of Fig. 22. If his squared paper is ruled in millimetre squares, he may find it convenient to take 4 centimetres as the wavelength of the lower note. He will then set out the sines of angles from 0° to 90° over the first centimetre, and so on. The wavelength of the upper note will be $2\frac{2}{3}$ centimetres; and all the distances from left to right for this note will be $\frac{2}{3}$ of the corresponding distances of the lower one. Having drawn both curves, he may calculate total displacements at a number of points by combining the separate displacements algebraically. He can then draw a thick freehand line through the points of total displacement, when he will obtain a curve like the thick line in (i) of Fig. 22. So that it can be seen clearly this thick line has been copied separately in (ii) of Fig. 22.

Now this thick wavy line represents the actual displacements of a succession of particles which occur simultaneously in the air when disturbed by two pure tones a perfect fifth apart. What happens when the wave made up of these displacements reaches the ear?

It is evident that the curve shown in (ii) of Fig. 22 is a periodic curve of wave-length equal to AC. It repeats over the distance CE with exactly the same shape, and it would go on repeating over equal distances to the right. Call its wave-length λ. Now the wave-length of the lower pure tone in our fifth is AB. The curve repeats over the distances BC, CD, and DE. Its wave-length would be $\dfrac{\lambda}{2}$. Similarly the wave-length of the upper pure tone in our fifth is one-third of the distance AC, or $\dfrac{\lambda}{3}$. It repeats over distances equal to one-third of AC. The question is, what sensation is produced in the ear by this periodic wave of wave-length λ, which is built up from two simple pendular vibrations with wave-lengths $\dfrac{\lambda}{2}$ and $\dfrac{\lambda}{3}$ respectively?

First of all we must examine more closely the progression of this wave through the air. In Fig. 20 the positions of the particles shown in (iv) were those which successive particles occupied at the same moment when disturbed by a pendular vibration. We saw that, at this moment, each particle was $\dfrac{1}{16}$ of a phase ahead of the next particle represented on its right. Thus, after a period equivalent to $\dfrac{1}{16}$ of a phase, particle 2 will occupy a position corresponding to that which 1 now occupies, particle 3 one corresponding to that now occupied by 2, and so on. Finally a new particle, which would have to be numbered 18, would occupy a position corresponding to that now occupied by 17. If, after this interval of time, we again drew the associated displacement curve it would have exactly the same shape as before, that is, as shown in (v), but would be moved bodily to the right by a distance equal to the distance between the equilibrium positions of particles 1 and 2. There would be no displacement for 2, maximum downward displacement for 6, no displacement for 10, maximum upward displacement for 14, and no displacement for 18. After a further equal period of time the whole curve would have moved to the right by another sixteenth of a wave-length.

Exactly the same thing happens to a more complex wave form like that in (ii) of Fig. 22. If an associated displacement curve such as this could be made visible, by some effort of imagination, we should see it moving rapidly to the right at a speed equal to the velocity of sound in air, but keeping its shape unchanged. The motion would appear, in fact, to be just what we should see if a piece of wire were bent into the shape of the wavy curve in (ii) of Fig. 22, made rigid, inserted into a piece of rubber tubing, and pulled through it at the velocity of sound in air. Each little piece of the rubber tubing would waggle up and down with the sort of motion with which, on a greatly reduced scale, the particles of air would swing backwards and forwards along the path of the sound.

Consider now the analogy used by Helmholtz, to which we referred in Chapter V, of sympathetic vibrations excited in the strings of a piano when the loud pedal was held down and a note was sung into them. Suppose that the sound represented by its associated displacement curve in (ii) of Fig. 22, was exciting sympathetic vibrations in the strings of the piano. The wave-length of this sound we have called λ. It would impinge on a string capable of sounding a note of wave-length λ. If this string were sounding its fundamental it would produce in the air a vibration with an associated displacement curve like that shown in (iii) of Fig. 22. We have drawn this immediately under the curve in (ii), because the marked difference in the shapes of the two curves suggests that the wave whose associated displacement curve is drawn in (ii) could not excite, in the string under consideration, a vibration which would cause it to emit vibrations into the air represented by (iii). In fact, it would not and could not excite such a vibration. But the vibrations shown in (ii) could and would excite the appropriate vibrations in pianoforte strings whose fundamental tones had wave-lengths equal to $\frac{\lambda}{2}$ and $\frac{\lambda}{3}$, that is, the notes an octave and a twelfth respectively above the first string we considered. The student, with the help of another, can try the experiment. Let them sing together two notes a fifth apart into the piano with the loud pedal held down, and then stop suddenly. The strings will go on sounding the same two notes first excited by sympathetic vibration.

The pianoforte strings are, of course, only an imperfect analogy for what happens in the ear; but they give a sort of picture of what probably does happen. As the ear receives a complex vibration like that in (ii), it analyses it into its simple component vibrations and hears those. It may be supposed that the complex vibration shown in (ii) excites, in the nervous appendages in the ear, sensations which are identical with those which would be excited separately by the simple vibrations of the pure tones they tell the brain they hear. Whatever pendular vibrations are put into a complex one, those vibrations are heard as notes by the ear; and the ear can get nothing out of a complex vibration except the pendular vibrations put into it.

This property of the ear, of analysing immediately and with certainty any complex vibration which is produced by combining two or more pendular vibrations, and thus, as it were, of performing calculations which mathematicians can make in the manner described in Appendix V, is a very wonderful thing. On it depends the whole possibility of music. Compare it with the behaviour of the eye. If light composed of different colours, that is, a vibration compounded of a number of vibrations of different frequencies, and therefore different wave-lengths, reaches the eye, it is not analysed by the eye into its constituent colours. The eye merely sees white or whitish light. Observe that, as stated above, the ear cannot get out of a complex vibration any simple pendular vibrations except those which were put into it. This, in effect, is what is stated by Ohm's law, referred to in Chapter IV, namely, that the ear can hear a particular tone in a sound wave which reaches it, only if a pendular vibration corresponding to that tone (including, of course, a component of a complex vibration) actually exists in the air. A very interesting passage in Helmholtz's work deals with this matter experimentally and more fully.

Now let us proceed to draw in the same way the associated displacement curves for two pure tones a major third apart. The ratio of the frequencies is $5:4$ and the ratio of their wave-lengths will therefore be $4:5$. Such curves are drawn in Fig. 23 and are completed until the point is reached at which the two curves are again in phase at the same stage of vibration as at the beginning. The distance covered will be five times the wave-length of the upper note and four times that of the lower note. If this distance is λ, the wave-length of the upper note would be $\dfrac{\lambda}{5}$, that of the lower note $\dfrac{\lambda}{4}$. The compounded vibration would be represented by a wavy line representing a wave with a wavelength of λ. It has been copied below so as to make its shape

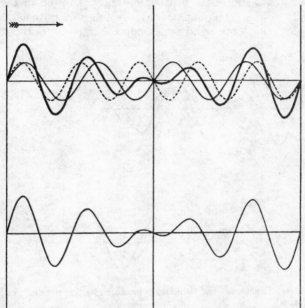

FIG. 23. Diagram of the sound wave produced by two pure tones a major third apart.

more evident. Its shape is not unlike that of Fig. 22 but it contains more waggles and the waggles in the middle will be flatter. When this compounded wave of wave-length λ reaches the ear, it is analysed into the two harmonic vibrations from which it was built up, and the ear hears two pure tones a major third apart and nothing else.

We can draw in the same way the compounded vibration representing the sound of two pure tones a minor tone apart (the two meanings of tone will not be confused). The compounded vibration, Fig. 24, which shows how the particles of the air actually move, is a wavy line similar to that produced by a fifth and a major third; but it contains still more waggles, and the flatter portion in the middle is still more conspicuous. But again the ear analyses it, and the only notes it hears are the two pure tones a minor tone apart.

In Chapter IV, when we discussed beats and the sensation of beating, it was stated that whether there was a sensation of beating or not, the physical beat existed between the two notes. That physical beat is represented in all these diagrams. It appears as a successive contraction and expansion of the amplitude of the vibrations. Physically this means that there is a successive decrease and increase in the intensity of the sound.

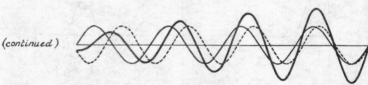

(continued)

FIG. 24. Diagram of the sound wave produced by two pure tones a 'minor tone' apart.

This successive decrease and increase in the intensity of the sound is the physical beat. If, as in Fig. 24, the beat is not too rapid it produces a sensation of beating in the ear, which for notes in the treble clef reaches its maximum unpleasantness at the interval of a semitone. There are, therefore, three things to distinguish. The first is the physical beat, of which we have drawn diagrams. The second is the sensation of beating, which does not occur till the beat is sufficiently slow to irritate the nerves of the ear. The third is the combination tone, which does not appear at all in these diagrams. Observe that, although no combination tone appears, beats which are too rapid to be audible appear in Figs. 22 and 23. Each of the tones we have combined in these diagrams is represented by a sine curve. Each is therefore due to a simple pendular vibration; and, as shown mathematically in Appendix III, this means that the tones cannot be loud. If they were loud, neither of them could be represented by a pure sine curve: something would have to be added to our diagrams; and when it had been added to, say, (ii) of Fig. 22, the ear would analyse the new combined wave, and say that it heard the two tones which we have already represented in (i) and also a deeper and much fainter tone, represented by a sine curve like that in (iii) but with a much smaller amplitude.

If this all appears fairly obvious it is because in the curves we have drawn we have made confident use of the knowledge which Helmholtz first made available. In Chapter VIII we shall give some indication of the really beautiful experiments he made concerning the quality of a note. Both by analysing complex notes and by building them up from pure tones, by his experimental technique, he made the acoustical basis for the quality of a note quite clear. Before his time it had been supposed that quality depended on the shape of the wave, but more by a process of reasoned elimination than anything else. The three physical data which characterize a vibration are its frequency, its amplitude, and its wave form. As frequency determined the pitch, and amplitude then determined the intensity of the sound, there was nothing left but the wave form to determine quality. Helmholtz was the first to demonstrate, as he did beyond the possibility of doubt, that the quality of a musical note depends only on the number, the selection, and the relative strengths of the partial tones of which it is composed. He also showed that the quality was independent of differences of phase between the constituent tones (see Chapter VIII, p. 103).

Let us then consider what happens if a pure tone is compounded with its first and second harmonic overtones, the octave and the twelfth, the one with half the wave-length and the other with a third the wave-length of the fundamental. Let us first combine graphically the sine curve of the fundamental with that of its octave, and let us suppose that the octave has two-thirds the amplitude of the fundamental. The result will be a curve similar to (i) of Fig. 25a, in which, as usual, the sound is travelling from left to right as shown by the arrow. Now let us copy on squared paper the wave form so obtained, and combine it with the sine curve of the second harmonic, giving to this half the amplitude of the fundamental. The result will be as shown in (ii) of Fig. 25b. The compounded curve is copied in (iii) so that its form may be the more clearly seen.

As a further example, let us combine a fundamental tone with its first overtone, assuming the amplitude of the overtone to be one-fifth that of the fundamental. The result will be as shown in (iv) of Fig. 25c; and, as before, the compounded curve is copied separately in (v). To keep the curves distinct in (iv), we have drawn the fundamental in (iv) with only two-thirds the amplitude of the fundamental in (i). In other words, the fundamental in (iv) is not so loud as in (i). It will be observed that in (iv) we have also represented the overtone as being completely out of step with the fundamental when the upward displacement of the fundamental is a maximum; and completely in step with the fundamental, where they are marked ✕, when the downward displacement of the fundamental is a maximum, whereas in both (i) and (ii) the components were all in step at a point of no displacement, marked ✕. Helmholtz's experiments tell us that this difference of phase makes no difference to the quality of the note.

We may illustrate, graphically, this conclusion of Helmholtz's experiments about differences of phase. Let us draw, as in (vi) of Fig. 25c, the fundamental and the octave harmonic with the same amplitudes as we gave to them in (i), but with the octave harmonic, or first overtone, completely in step with the fundamental when the upward displacement of the fundamental is a maximum. Both are marked ✕ at this point. As the curves we have drawn are associated curves, this means that the fundamental and the first overtone are both impressing their maximum forward displacement on the same particle of air at the same time. The effect is to add the displacements. It is evident from the component curves that when the fundamental is impressing its maximum backward displacement on a particle of air, represented by a downward displacement in our curve, the overtone is impressing its maximum forward displacement on

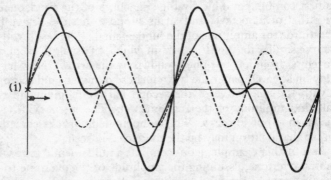

Fig. 25a. Diagram of the sound wave produced by a fundamental and its octave harmonic.

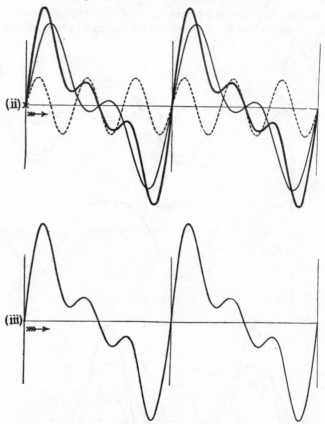

FIG. 25*b*. Diagram of the sound wave produced by a fundamental and its first two harmonics.

the same particle. The actual displacement of the particle is the difference of these two displacements.

We have now two curves, the thick line in (i) and that in (vi), which represent the combination of the fundamental and the first overtone without alteration of amplitude but with an alteration of phase. Their appearance to the eye is quite different; but Helmholtz's experiments tell us that the ear cannot distinguish between the sounds they represent. In each case the ear would hear a loud fundamental and a very loud overtone.[1] If we draw the curves again but with the overtone completely in step with the fundamental when the downward displacement

[1] See Appendix V.

of the fundamental is a maximum, and consequently completely out of step when the upward displacement of the fundamental

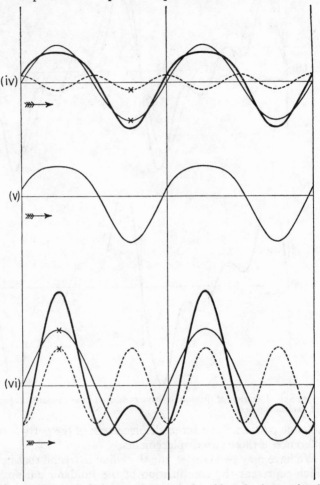

FIG. 25c. Diagrams of the sound wave produced by a fundamental and its octave harmonic, soft in (iv) and (v), loud in (vi) as in (i).

is a maximum, we shall find that when combined they give a thick line exactly like that in (vi) but turned upside down. Or by merely turning the book upside down we can observe the alteration of phase we should obtain. The ear would not be able to tell the difference, for it is caused only by an alteration of phase.

The thick line in (iv) turned upside down would resemble that in (vi) but without the waggle in the flat portion. This waggle is, of course, the result of representing the overtone as being much louder in (vi) than it is in (iv). The ear could tell this difference. It would say that the note in (iv) was not so loud as in (vi) and that its quality was different, for the overtone would be much fainter.

The results shown in (iii) of Fig. 25b and (v) of Fig. 25c are very interesting. If the student can consult in his college library the *Dictionary of Applied Physics*, edited by the late Sir Richard Glazebrook, he will find in an article on Sound by Barton, a curve actually recorded from the sound of a flute by Miller's phonodeik. This was an apparatus in which sound caused a membrane to vibrate, and its vibrations were recorded optically on a strip of photographic film moving rapidly, as a cinematograph film is made to move. Traces made by the phonodeik when the flute was played forte and when it was played piano appear from the reproduction in the *Dictionary of Applied Physics* to have been like the curves we have shown in (iii) of Fig. 25b and (v) of Fig. 25c. The amplitudes used in (iv) of Fig. 25c for the component sine curves were chosen to keep the three curves of the figure distinct. The resemblance between (v) of Fig. 25c and the trace made by the phonodeik of the sound of a flute played piano would be closer if the fundamental in (iv) were drawn with a much larger amplitude, and the octave harmonic with a smaller one. This illustrates the observed fact that most of the loudness of a flute played forte is due to the addition of powerful overtones.[1] This fact has an analogy in the organ. The addition of the principal, fifteenth, and mixtures makes the diapason sound much louder, as well as more brilliant.

The softer notes of the flute consist of almost pure tones. They therefore lack definition, and if played as a solo instrument the flute becomes rather monotonous. Helmholtz quotes an old joke that, for this reason, the only thing more dreadful to a musical ear than a concerto for the flute is a concerto for two flutes. The note of the oboe, on the other hand, is very rich in upper partial tones. This gives to its wave form a highly serrated appearance. In the article in the *Dictionary of Applied Physics* there is reproduced a trace, made on the phonodeik, of the sound of an oboe which exhibits such an appearance.

Throughout this chapter we have spoken of particles of air as

[1] See p. 123 and Appendix V.

if they had a concrete existence. The student must not imagine that they refer to anything in the structure of the air like the molecules. A particle may be imagined as a small volume of air, say spherical in shape, which is so small that its shape and size are not affected by the movements to which it is subjected.

VII

SOME PROPERTIES OF SOUND, AUDIBILITY,
SENSITIVENESS OF THE EAR

WE saw in the preceding chapter that sound is propagated through the air by vibrations which travel in progressive waves. Light is also propagated by vibrations, and we should therefore expect sound to exhibit properties of light such as reflection, refraction, and interference. In fact it does so; and text-books on sound give directions for the performance of experiments to demonstrate the fact, thus confirming the correctness of the presumption that sound does travel through the air in the way we have supposed. The place for experiments of this kind is the lecture-room or the laboratory, and we shall therefore not delay to describe them in this book. Those who wish to know how they can be performed may consult books on acoustics. We may merely remark on the difficulty, which arises from the great difference in their wave-lengths, of designing experiments to show that sound behaves like light. It would be possible to insert about two million wave-lengths of yellow (sodium) light in one wave-length of the sound of middle C. To bring many experiments on sound within the dimensions possible for practical trial it is therefore necessary to employ notes of very high frequency. Instruments such as Galton's whistle or Rayleigh's bird call have been designed for the purpose. This great difference between the wave-lengths of sound and light makes it impossible for small objects, which throw perfectly good light shadows, to form sound shadows. Even a house throws only an imperfect sound shadow. We may add that, while many of the laboratory experiments on sound have historical interest, such as experiments on organ pipes briefly indicated in Chapter IX and measurements of frequency or absolute pitch, the technique employed experimentally to-day in research work on sound has been revolutionized since Helmholtz's day by the use of electrical means; for example, the use of the thermionic valve which has given us broadcasting, or of the cathode-ray oscillograph which has given us television.

The best demonstration of the truth of the law of gravitation is to be found in the means it has afforded for calculating the motion of the planets. The most signal example was the discovery, by John Couch Adams, working at his desk—a discovery made almost simultaneously by the French mathematician Leverrier—of the planet Neptune from the disturbances it was causing, through gravitation, in the orbit of the then recently discovered planet Uranus. Similarly, the real demonstration of the truth of what we have supposed to be the nature of sound lies in the fact that all the phenomena to be deduced from our supposition occur in nature. This is particularly true, as we shall see in Chapter IX, of interference which, for example, furnishes explanations of the behaviour of organ pipes that agree with the observed facts.

We shall confine ourselves in this book to some observations on the properties of sound which have direct interest for musicians.

Reflection.

Reflection of sound is familiar to every one in the form of echoes. Echoes have a practical interest for musicians because they have important effects on the sound heard in enclosed spaces such as music-rooms or concert-halls. A brief note on the qualities of auditoriums will be found in Appendix XI.

The laboratory experiments referred to above are intended to show, amongst other things, that reflection of sound is essentially similar to reflection of light. But, if we are to understand effects of reflection of sound which are interesting to musicians, it is important to realize how reflection of sound differs, in degree, from reflection of light. When light falls on window glass some of it is reflected, but most of it enters and passes through the glass. When sound, travelling through the air, falls perpendicularly on a solid surface, a trifling amount enters the solid but nearly all of it is reflected. This is very interesting because sound travels very well in solids such as wood or metal, a fact made familiar to us all by the way in which sound can travel through a house in the water-pipes. In this sense wood and metal are transparent to sound. Yet they do not allow air-borne sound to enter them readily in the way that glass allows light to enter.

Hard painted plaster reflects sound very well. If sound falls perpendicularly on such plaster, say in the wall of a concert-hall, all but one or two per cent. of it is reflected. A formula due to Rayleigh[1] shows that if sound of the frequency of middle C were to fall perpendicularly on a sheet of iron half an inch thick and

---

[1]   *Theory of Sound*, 2nd ed., § 271.

rigidly fixed, all but one part in 50,000 would be reflected.[1] All but one part in 200,000 of sound an octave higher would be reflected. If the iron sheet were free to vibrate, rather less of the sound would be reflected: a small part of the energy of the vibrations in the air would be absorbed in causing the iron sheet to vibrate. The thinner the sheet the more readily would it be set in vibration in this way. The vibrating sheet would emit sound waves from the opposite face: the sound might be too faint to be audible, but if not it would give the listener the impression, wrongly, that it had all passed through the sheet of metal in the way that light passes through a sheet of glass.

The important conclusion for musicians is that solid surfaces are extraordinarily good reflectors of sound. The complete explanation of this phenomenon is beyond the scope of a book of this kind. But, if what we have already noted of the relative sizes of a wave-length of sound and one of light be recalled, it is evident that argument by analogy may be particularly dangerous in this case. We shall avoid misconception if we always regard the pressure of the air at any point as the characteristic which determines its behaviour there when disturbed by a sound wave. A condensation represents something added to the normal pressure. A rarefaction represents something subtracted from it. The oscillations of our supposed particles of air should always be pictured as the effect of which the changes of pressure are the cause. We shall be interpreting correctly the results of both mathematical calculation and laboratory experiment if we think of a sound vibration travelling in a light medium like air as being unable to cause an appreciable degree of sound vibration in a dense and heavy medium like metal, even if its line of progression be perpendicular to the surface of the metal. We can picture this as the reason why condensations are reflected there almost completely, and a similar picture would represent the behaviour of rarefactions. On theoretical grounds, explained in books on sound, a condensation or rarefaction in a sound wave in air which reaches a solid surface at an oblique angle would be totally reflected there. But the distinction between a perpendicular and an oblique reflection is of little practical moment, for it hardly ever happens that the incidence of sound on a solid surface is wholly perpendicular.

Moreover, if the solid surface were that of a thin sheet of metal, a condensation or rarefaction reaching it obliquely or

[1] In this theoretical deduction it is assumed that iron is a perfectly non-absorbing material, which is not quite true.

even tangentially could excite some vibration in the sheet, which in turn would emit sound waves from its other side. A condensation would tend to press it outwards, a rarefaction would tend to suck it inwards. Thus the thickness, density, and rigidity of the solid are the really important factors which decide how completely sound shall be reflected by it. It is common knowledge that a sufficiently thin and non-rigid partition will allow airborne sounds on one side of it to be heard on the other side of it, while a thick solid wall will be sound proof.

The preceding observations apply to a non-porous solid surface. If the solid surface is porous, it is able to absorb sound. Sound is a form of energy; and in the interstices of a porous body it is rapidly converted into heat by forces of a frictional nature. As sound, it disappears. This is important in music-rooms and concert-halls, as explained in Appendix XI.

Since a metal sheet can readily reflect sound and so prevent most of the sound from being transmitted to its other side, we may expect that when it takes the form of a tube it can be used to convey sound. As is well known, sound can be carried to a distance with little loss in a speaking-tube; and it can be taken in the tube round corners by reflection at the internal surface of the tube. Whether the tube be straight or bent, no sound will escape from it unless its walls are free to vibrate. In that case its walls will be set in vibration by the condensations and rarefactions which pass along them; and, if the walls are thin enough, faint sound waves will then be emitted from the outside surface of the tube.

Every musician knows that while a post-horn, such as was used on coaches, is straight, the horn used in the orchestra is bent into a circle. The bassoon is essentially a large oboe, but if it were constructed like a large oboe with a straight tube its length would make it unmanageable: the tube is therefore bent back upon itself with no ill effect. From what we have learnt above about reflection of sound, we should expect that it would not matter to sound waves travelling along the tubes of these instruments whether the tubes were straight or bent. But the effect of bending in the tubes of wind instruments is not quite the same as that of bending in speaking-tubes. When we come to Chapter IX we shall find that the movement of the air in organ pipes, or the tubes of wind instruments, is different from the movement of the air in a speaking-tube along which sound waves are passing; but that there is a close analogy will then be evident (see p. 124). We shall also meet, then, in the stopped pipe of the organ, an example of direct reflection of condensations and rarefactions in which there is no appreciable loss, for reasons explained in this chapter.

Interference.

Interference results when the troughs of one sound wave combine with the crests of another sound wave to oppose each other. We drew curves in Chapter VI to show what happened when waves became out of step and produced beats. Obviously if two sine curves are drawn to represent the associated displacement curves of two sounds of the same frequency and the same amplitude and completely in step, they will reinforce each other and produce a sine curve with twice the amplitude of that of the component curves. But if they are completely out of step they will cancel each other, as in Fig. 26, and no sound will be heard.

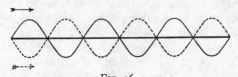

FIG. 26

Their combined curve will be the thick straight line, which represents a wave with no amplitude at all, that is, no sound at all. No displacement and no condensation of the particles of air occurs.

The student can hear an example of this if he listens to a tuning-fork and, while holding it close to his ear, turns it in his fingers. Four times in each complete turn the sound will disappear. In Chapter VI we considered the sound from a tuning-fork and we concentrated on the effect of one vibrating surface. Actually there are four, the inside and the outside of each prong. Their combined effect is that, in four directions at angles of about 45° with the plane in which the prongs are vibrating, the condensations from two surfaces always cancel the rarefactions from the other two and vice versa. In Chapter IX we shall discover the effect of interference between two sounds such as those whose displacement curves are shown in Fig. 26, but travelling in opposite directions.

Audibility.

The analogy between light and sound suggests inquiry into their analogous effects on the eye and the ear, of which we have already had an example when we compared the sensation of beating with that of flicker. The eye sees light of different colours in the spectrum of which that of highest frequency, violet, vibrates about twice as rapidly as that of lowest frequency, red. The frequency of visible light therefore extends over an octave. But there is light of higher and lower frequency which can be seen by the photographic camera. The film supplied for photographic work is sensitive to what is called ultraviolet light, that is to light whose frequency is greater than that of light at the violet end of the spectrum. Recently photographic material has been invented which is sensitive to light of lower frequency than that at the red end of the spectrum; and it has made possible what is known as photography by infra-red rays.

Exactly analogous effects occur with sound and the ear. The maximum range of frequency of sounds which are audible varies between individuals, but may be taken as extending from a frequency of 20 to one of 20,000. In older people the limit of audible sound is less. The squeak of a bat can be heard by most young persons: to older ones it is often inaudible. The ear of young people can thus hear a range of sound which extends over about 10 octaves. There are sounds which have lower and higher frequencies than those within the limits given, but we cannot hear them.

But the ear is not equally sensitive to all audible sounds. It is most sensitive to sounds whose frequencies lie between 500 and 5,000. These are approximately the frequencies of C in the treble clef and of a note three octaves and a third above it. The ear is less and less sensitive to sounds whose frequencies become progressively lower than 500. To become audible, notes of frequencies less than 500 therefore require to have increasing intensities as they become deeper in pitch.[1] Finally the power of the ear to hear the note at all disappears, but lower notes such as the fundamental tones of a 32-foot stop on the organ can be felt. Above the upper limit the ear rapidly becomes less sensitive to sounds, and if musical notes of these higher frequencies are to be audible they must have greater intensities than those required if notes of frequencies between 500 and 5,000 are to be rendered audible. Thus, beginning at the bottom of the scale, the intensities which are required by sounds to enable the ear to hear them diminish as we rise up the scale, till we reach notes near the top of the treble clef. For very high notes the intensity has to be increased again to render them audible. Finally, sound

[1] This has an appreciable effect on the audibility of low difference tones.

of even the utmost intensity is inaudible. The range of intensities at which sounds of different pitch just become audible is called the threshold of audibility.

We have therefore to distinguish between intensity of sound, which is a physical measurement,[1] and the loudness of sounds as heard by the ear. To produce over the whole keyboard sounds whose degrees of loudness balance, it is necessary to give to the lower notes of the piano and the organ an intensity which is out of proportion to their loudness. Makers of these instruments have to allow for this. For the same reason, if an instrument which measured sounds by their intensities could be used to listen to an orchestra, the bass instruments would be altogether overpowering.

If notes in the region in which the ear is sensitive become exceedingly loud they also become painful. They are then said to reach the threshold of feeling. In the regions in which the ear is most sensitive, the intensity of sound on the threshold of audibility has to be increased about ten million million times before it becomes painfully loud and reaches the threshold of feeling. In the bass and for very high notes the intensity required before the sound is felt is much less. Very shrill notes may easily become painful, and some people feel the lower notes of the 32-foot stop on the organ so acutely as to find them painful.

Sensitiveness of the Ear for Pitch.

The ear finds it difficult to distinguish the pitches of very high notes or very low notes. Most people are familiar with the difficulty of placing exactly the pitch of two or three notes at the bottom of the pianoforte keyboard, although they are reinforced by the octave harmonic. In the regions in which the ear is more sensitive, very small differences of pitch can be discerned by the ear. But if the difference is too small, the ear hears only a single note in which there are audible beats at a rate equal to the difference in the frequencies. The note which the ear thinks it hears lies between the two notes actually heard. This is familiar to organists. If after using the voix celeste stop with its companion stop, to which it is tuned sharp, the organist pushes the celeste in, the pitch sounds as if it had been very slightly flattened.

Sensitiveness of the Ear for Overtones.

The ear's power of blending the sound of the upper partials with the fundamental tone, and calling the result a single note

[1] See Appendix V.

with a characteristic quality, is particularly interesting to musicians. The ear is able to some extent to distinguish overtones only if it wants to do so. Some idea of this can be obtained by training it to pick out the harmonic overtones in a note. Interesting experiments can be tried on the pianoforte. Strike C in the bass clef loudly and, holding the digital down, touch lightly and for a moment one of the notes of its harmonic series. The ear will continue to hear the same note in the overtone of the bass note. By intent listening after the bass note is again struck, the ear will be able to isolate the overtone without asking for its note to be touched lightly first. The sustained note of the violin or the organ is still better for experiment. As the ear, with practice, learns to concentrate on the exercise of this power, it is surprising to find how the sound of the overtone appears to increase in volume. There can be little doubt, as Rayleigh observes,[1] that the difficulty which the ear experiences in detecting the overtones of the human voice is partly the result of trained habit. In this sense one might say that the ear ignores the overtones, not just because it does not want to detect them, but because it wants *not* to do so. The description of an experiment which enforces this will be found in Helmholtz's book. In a most interesting passage Helmholtz discusses all this as a question of physiology, and draws the necessary distinction between the sensation which the ear excites in the nervous system, and the perception of the sensation by the brain. And Raleigh remarks: 'Most probably the power of attending to the important and ignoring the unimportant part of our sensations is to a great extent inherited—to how great an extent we shall perhaps never know.'

Interesting light is thrown on the ear's power of blending the partial tones with the fundamental by what it makes of the note of a church bell. Most of the overtones of a bell of haphazard design would be inharmonic. The art of the bell-founder lies in making some of the overtones harmonic or as nearly so as possible, whereby the note of the bell becomes more sweet and less of a jangle. The deepest tone of the bell, and the most persistent one, is called the hum. The first overtone is approximately an octave higher and is called the fundamental. The next overtone is called the tierce, and in a good bell it may be either a major or a minor third above the fundamental. In modern practice in this country the minor third is preferred by the designer and tuner of the bell. Tuning is done by placing

[1] *Theory of Sound*, 2nd ed., § 25.

the inverted bell below a machine with a rotating cutting tool; and its art consists in removing metal from the right place inside the bell. The next two overtones are the quint, about a fifth above the fundamental in a good bell, and the nominal about an octave above it. It is the nominal which gives the name to the pitch of the bell in a peal. When not a perfect octave above the fundamental, it appears to have the remarkable power of leading the ear to imagine that the tone of the fundamental is an octave below that of the nominal. The 'striking note', which the ear says it hears an octave below the nominal, is thought to be an aural illusion.

VIII

RESONANCE, FORCED VIBRATIONS, RESONATORS, HELMHOLTZ'S EXPERIMENTS

WE have still to make acquaintance with one important acoustical effect which contributes to the sound of musical instruments. A piano without a soundboard would be a futile instrument. A violin which consisted merely of a frame without belly or back would be a mere ghost. Instruments of this nature, in fact, have been made for violinists to use in exceptional circumstances in which it would not be possible for them to practise on the violin itself. Such instruments give only faint notes. Why does the soundboard make all the difference?

Let us turn to other vibrating bodies for information. If a child's swing is swinging at the full, it will gradually come to rest if the person swinging it applies no further pushes to maintain the oscillations. This gradual destruction of the motion of the swing is, of course, due to the resistance which the air constantly opposes to it, to the friction of the air slipping past it, and to that of the loop and hook at the cross-beam. If a series of quite gentle pushes be now given to the swing, so as to synchronize with the natural period of its oscillations, it will swing gradually further and further until it is again oscillating at the desired limit. But if the person pushing the swing mistimes the pushes, the swing, though set in motion, will not get very far. The occupants of a swing-boat at a fair set it in motion by pulling alternately on two ropes attached to either end of a rigid bar which is laid at right angles to the cross-beam. If they persist in pulling too quickly, a motion will be imparted to the swing which will have to keep step with their pulling, but the swing will oscillate with an amplitude which is relatively small.

These motions are characteristic of the behaviour of all vibrating bodies. If a vibrating body is left to itself, its motion is gradually destroyed by dissipation of its energy through the formation of waves or the resistance of forces of a frictional nature. The formation of waves accounts for a large part of the energy required to drive a ship. Stream-lining is important in the design of anything which has to move rapidly through air because it reduces the energy which is absorbed in making waves in the air. The loss of energy in a vibrating body through the creation of waves, which in suitable circumstances may be sound waves, can be an important factor in reducing and eventually stopping its vibrations. The destruction of the motion of a vibrating body in this way is important in the mathematical analysis of vibrations, and is known as *damping*. If the damping is slight the body will vibrate for some time. If the damping is considerable it will soon be brought to a standstill. The rate at which the child's swing is brought to a standstill by forces that resist its motion is the measure of the damping of the swing.

When periodic impulses impressed on the motion of a body synchronize with the natural period of the vibration of the body, the amplitude of the vibrations eventually set up is usually considerable. We saw that this was true of the child's swing. Bridges have natural periods of oscillation; and, to avoid exciting oscillations which might become serious, soldiers on the march are often made to break step on crossing bridges. Recent researches into the stresses caused in railway bridges by engines and trains passing over them, showed that old-fashioned engines of relatively light weight might exercise more injurious effects on bridges than modern heavy engines, because the moving parts of the old-fashioned engine, being sometimes insufficiently balanced, could set up extraordinary vibrations in bridges if the period of their unbalanced oscillations happened to coincide with the natural period of vibration of the structure of the bridge. The creation of large vibrations in a body by relatively small periodic forces which are in step with the free vibration of the body, is called *resonance*. The sympathetic vibrations we considered on pp. 49 and 77 were examples of resonance.

When the periodic forces are not naturally in step with the free vibration, the vibrations ultimately set up are called *forced vibrations* if they keep in step with the periodic forces. Their amplitude is less, and often considerably less, than in the case of true resonance. Our illustration of the swing-boat made to swing unnaturally, by pulling on the two ropes in turn too rapidly, was an example of a forced vibration.

These general properties of vibrating bodies are true of bodies whose vibrations cause audible sound waves. The creation of the sound waves, by exhausting the energy of the vibrating body, causes the intensity of the sound to diminish and the sound ultimately to cease, unless the vibrations are maintained by some exciting force. The loudness of the notes sounded by the pianoforte diminish rapidly for this reason. The notes of the violin are sustained because the bow keeps seizing the string and imparting new energy to its vibrations. Those of wind

H

instruments, or of singers, are maintained by the breath of the player or singer expelled by muscular pressure. The notes of the organ are maintained by the energy employed in working the bellows.

When a string is vibrating, the air is pushed backwards and forwards by the vibrations only to a limited extent; much of it passes round the advancing string and closes in again behind it. The forced vibrations excited in the soundboard are required to impart the necessary loudness to the sound heard. A tuning-fork held in the hand produces an effect not unlike that of a string without a soundboard. Partly its vibrations cause compressions and rarefactions in the air, partly the air keeps slipping round the prongs as they vibrate. That is one reason why a resonance box is used with a tuning-fork, as we shall see.

It is very easy for the student to perform experiments for himself which will tell him more about resonance in half an hour than any amount of book description. All he needs are a tuning-fork, some short-necked bottles, and a jug of water. Take a short-necked bottle and blow over the mouth. If the bottle is suitable for his experiments it will give a clear note. Now pour water into the bottle and blow again. The note of the bottle will have risen in pitch. If the bottle is now tilted, so as to alter the shape of the portion occupied by the air, the pitch of the note produced by blowing over the mouth remains unchanged. The shape of this space therefore does not matter; what matters is its volume.

Now take two bottles which give good notes, but select one bottle as having a much narrower neck than the other. Fill the larger bottle to the bottom of the neck. From it fill the smaller bottle to the bottom of the neck without spilling any water. If necessary use a funnel. The volume of air in the larger bottle is equal to that of the water in the smaller one. Empty the smaller bottle and we now have two bottles containing equal volumes of air, one with a narrower neck than the other. Blow over their mouths. The one with the narrower neck has the lower note. Thus, if we increase the size of the opening into the bottle we raise the pitch. This information will be interesting when we come to discuss some of Helmholtz's experiments.

Now set the tuning-fork in vibration and place its stem on the table to make its fundamental note sound loudly. Take a bottle which gives a note of lower pitch than the tuning-fork when its mouth is blown over, and pour water into it till the pitch of the note sounded by blowing over it is as close to that of the note of the tuning-fork as possible. If the tuning-fork is now set in vibration and held over the mouth of the bottle a loud clear note will be heard. This note will have the same pitch as the note of the fork, or as the note sounded by the bottle when it is vibrating in its own natural free mode as the result of blowing across its mouth. We have made an example of *resonance*.

Now pour water out of the bottle and repeat the experiment. The same note will be heard but it will be fainter. Pour more water out and the note will be fainter still. Fill the bottle again to the point at which its natural note is the same as that of the fork. A loud clear note will again be heard when the vibrating fork is held over the mouth of the bottle. Pour in some more water. The note becomes fainter. Pour in still more water; the note becomes fainter still, but it will still be a note of the same pitch as that of the fork. When water is poured out or added, the natural note of the bottle is altered. As the bottle still vibrates to the pitch of a vibrating fork held over it, the vibrations excited in it by the fork are not its natural vibrations. We have made an example of *forced vibrations*. The fork being a strong structure, its vibrations are the dominant factor in the combined system of bottle and fork.

A combined system consisting of a source of sound and a means of enhancing the loudness of the sound by either reso-nance or forced vibrations is called a *coupled system*. It depends on the nature of a coupled system whether the pitch of the note produced by it is that of the source or that of the note natural to the connected part of the system which enhances the loud-ness of the sound. In the pianoforte and the violin the sound-board, as we know, has to vibrate exactly as the strings require it to vibrate. The string plus soundboard is a coupled system in which the note of the string is master; and the system has to sound the note of the string. The soundboard therefore makes forced vibrations. In an organ pipe or a wind instrument it is the pipe or the tube of the instrument which is master and deter-mines the note of the coupled system. The reed of a wind instrument is very flexible, and the tube of the instrument exhibits true resonance and makes the vibrations of the reed con-form to its own period of vibration. We shall discuss this again in the next chapter, when we shall find in the reed pipe of the organ a coupled system in which there is mutual constraint. Neither the reed nor the pipe can be master, and so unless their natural notes are made as close as possible, the sound of the coupled system is a failure. All musical instruments are examples of coupled systems.

A body which exhibits resonance is called a *resonator*, and its adaptability to notes of different pitch depends on its own way of vibrating. If the note of the resonator dies very rapidly the damping of the resonator is considerable. If it dies relatively slowly the damping is small. If the damping is small the resonator only sounds loudly when excited by notes whose pitch is close to its own. If the damping is considerable the resonator can respond to a wide range of notes of different pitches, though it still gives the best results with a note of the pitch of its own natural vibrations. It is not difficult to understand in a general way why this should be so. The first effect of the exciting sound on the resonator is to make it try to vibrate in its own free period. If the damping is considerable the amplitude of the vibration so caused rapidly diminishes as shown by the damped vibration drawn, from calculations, in Fig. 27.

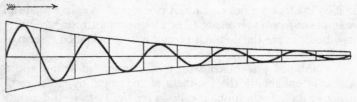

FIG. 27

Consequently, before another vibration of somewhat different wave-length imposed on the initial one can get seriously out of step with it, the original vibration is moribund and the new vibration has the mastery. It is not possible to show in detail, without mathematical analysis, exactly what happens; but it is easy to understand why the forced vibration meets with comparatively little resistance, though as the resonator keeps trying to start its own vibrations, so to speak, the forced vibration meets with enough resistance to prevent it from having an amplitude approaching that of the note sounded when there is true resonance. If however the damping is slight, the vibration which the resonator desires to make, and which is excited to begin with, soon gets seriously out of step with the new vibration of different wave-length: as the mastery between them is seriously disputed they destroy each other in the resonator more or less completely. A resonator whose damping is small is therefore very selective in the notes to which it will respond.

When held in the air a tuning-fork gives vibrations whose sound has little intensity. But if the stem is held on a table, the fork excites in it forced vibrations which greatly magnify the intensity of the sound. The vibration of the prongs of the fork cause the stem of the fork to move up and down very slightly. These movements are transferred to the table or other object against which the stem is pressed. If the object has a large surface it transmits a loud sound to the air, acting as a soundboard.

Tuning-forks for experimental use are fitted to resonance boxes which are shaped very much like a large match-box with one end removed; and on their top they have a small knob with a hole in it into which the stem of the fork can be inserted. Reference was made to a resonance box for a tuning-fork in Chapter IV. A resonance box may be compared to a pipe closed at one end; and the air in it has a natural period of vibration. The best result is obtained when the air in the resonator has the same fundamental vibration as the tuning-fork; but the damping of the resonance box is considerable, and the same box can be used with forks of somewhat different pitch. It always sounds the note of the fork. On the other hand, the air in a resonance box with the same fundamental as the fork, resents the inharmonic overtones of the fork, because they try to impose on it forced vibrations far removed from its own fundamental tone and out of step with what might be its own partial tones. It has no difficulty in rejecting the inharmonic overtones, and it resounds powerfully with the fundamental tone of the fork.

The coupled system of fork and resonance box makes a much louder tone than the fork alone; this means that, in making waves of sound, it uses up the energy of the fork's vibration more rapidly than does the fork by itself. The sound of the fork, by itself, consequently persists much longer than the sound of the coupled system.

We have already had occasion to mention the inharmonic overtones of the tuning-fork. The first overtone is usually in the neighbourhood of two octaves and a major sixth above the fundamental, but it varies a little with different forks.[1] Each prong of a fork is like a rod fixed at one end. The vibrations of a rod fixed at one end are interesting not only in connexion with tuning-forks

[1] When sounded very vigorously tuning-forks produce secondary overtones, including an octave harmonic which in some forks is quite pronounced. Rayleigh noted the difficulty this presented for fine work when a *loud* pure tone was needed. The conditions in the fork which produce secondary overtones are in some ways similar to those which produce combination tones in the ear (see p. 40) when notes are loud. For this reason Helmholtz compared the octave harmonic to a summation tone which the fundamental vibration of the fork produces with itself.

but also in connexion with vibrating reeds fixed at one end. They are therefore dealt with at length, in books on acoustics. The student can learn all he wishes to know about them if he carries out the simple experiment described in Appendix VI.

The effect of damping on the selective power of a resonator has been discussed at some length, because it is important in comparing with our bottles the design of the resonators used by Helmholtz in his experiments. His resonators had practically no necks; and for reasons explained in books on acoustics the result was to make them more selective than our bottles. Unless this be explained their difference from our bottles as resonators would be rather puzzling. Helmholtz's resonators were of two main types. One type was in the shape of a sphere and it had two orifices, one corresponding to the neck of our bottle, the other, opposite to it, being drawn out into a short tube which he inserted in the opening of his ear. To quote his own words: 'If we stop one ear, and apply a resonator to the other, most of the tones produced in the surrounding air will be considerably damped; but if the proper tone of the resonator is sounded, it brays into the ear most powerfully.' Helmholtz used resonators of this type to analyse complex notes. He had a series of them with which he could pick out and identify particular tones, such as the higher partial tones. By this means he was able to establish experimentally many of the results mentioned in the second half of this book. When one considers the simple character of his equipment one cannot but admire the experimental skill which enabled him to obtain such complete and comprehensive results.

As a complement to these experiments, and as affording the final proof of the soundness of his conclusions, Helmholtz devised another type of resonator, cylindrical in shape, with an orifice at one end which could be closed by a sliding cover. He made a series of eight such resonators each sounding as its natural note a note of the harmonic series, thus giving him the first eight partial tones of a complex note. In front of their orifices he fixed tuning-forks which sounded notes of the same harmonic series. Each fork with its resonator formed a coupled system which gave a pure tone, like a tuning-fork held over a bottle with the same natural note as the fork. The tuning-forks could all be set vibrating by electro-magnets along which intermittent currents were passed at a frequency equal to that of the fundamental tone of the series. This was accomplished by a device, in principle not unlike the make-and-break in an electric bell, operated in another room (to avoid disturbance from its sound) by another tuning-fork in unison with the lowest of the eight forks. He began an experiment with the slides closing the orifices of all his resonators. The apparatus then emitted only a gentle hum. He then opened the orifices of those resonators which gave him the partials he desired, including the fundamental. He reduced the loudness of any given resonator when he wished by drawing it away from its fork. By this means he was able to build up from its constituents the characteristic sounds of organ stops. The high upper partials required to imitate reed instruments were, of course, lacking; but he reproduced the nasal tone[1] of the clarinet by using unevenly numbered partials and the softer tones of the horn by the full chorus of all the forks.

In Chapter VI we mentioned that Helmholtz showed by his experiments that alterations in the phase of partials did not affect the quality of the complex note. It can be shown mathematically that the forced vibration which a fork causes in a resonator tuned to a pitch below that of the fork, differs in phase from the vibration caused when the fork and the resonator are in tune. To lower the pitch of the resonator Helmholtz partially closed the orifice with the sliding cover. As we have seen, if two resonators of equal volume have orifices of different sizes, the one with the smaller orifice sounds a lower note than the other. When he had obtained the quality of note he desired he readjusted the resonator, whose orifice he had partially closed, by opening the cover to the full and moving the resonator a corresponding distance from the fork to maintain the same intensity of tone. The result was to show in repeated experiments that the adjustment made no difference to the quality of the note heard from all the forks concerned.[2]

[1] The ambiguities in the use of the word 'tone' are unfortunate, but unavoidable in a book such as this. The musician uses the term in two senses: first, as a musical interval, the major or minor tone and the semitone; second, to describe the quality of a musical note, as in brilliant tone, round tone, reedy tone, diapason tone, or tone production. Scientific writers add two more meanings: a third, as describing an element in a musical note, as in prime tone, difference tone, or overtone, when they use it as meaning a pure tone; and a fourth, as describing a sound of recognized pitch, when it represents a complex of pure tones (see p. 7). The context will make the meanings in this book clear to the musical reader.

[2] Some other investigators have regarded this conclusion as only approximately true. Even if they were right, which may be doubted, the distinction would be of no practical importance for our purpose. Rayleigh concluded that the departure from Ohm's law, which would be involved in any recognition of phase relations, could only be ascribed to secondary causes. (*Theory of Sound*, 2nd ed., § 396.) More recent investigations, in which electrical means were used, tend to support Helmholtz's conclusions.

Reference to Helmholtz's experiments would be seriously incomplete without some mention of his investigation of vowel sounds. Helmholtz's experiments proved conclusively that the sounds of different vowels are determined by a suitable combination of partial tones. The student who desires to pursue this subject should turn to the pages of Helmholtz's own work.[1] Helmholtz's results have been confirmed and expanded by more recent researches, notably those of Sir Richard Paget recorded in the *Royal Society Proceedings*, 1923, 1924, and 1927, as well as by those of J. Q. Stewart published in *Nature*, vol. 110, 1922. It may be useful to mention one point of interest to musicians. The sound of the vowel *e* contains very high partial tones. In a building in which there is considerable reverberation, these high partial tones are damped, or absorbed, more rapidly than the fundamental tone. Before the sound dies away it takes on the character of an *oo* sound, for this reason. This illustrates the reason why the vowels sung by a good choir in a cathedral are often so difficult to distinguish. The fault does not lie with the choir.

---

[1] See also *Theory of Sound*, Rayleigh, 2nd ed., § 397, for an account of pioneer work in this field by Willis (*Camb. Phil. Trans.* 1829) to which Helmholtz makes reference.

IX

STATIONARY WAVES, THE ORGAN PIPE, ORCHESTRAL WIND INSTRUMENTS

Stationary Waves.

IN Chapter VI we drew diagrams of the vibrations caused in air by sound proceeding in a given direction. Waves such as those then produced are called progressive waves, for we showed that after an interval of time the wave form would still have the same shape but would be moved a corresponding distance in the direction in which the sound is travelling. We also drew the wave forms of complex notes produced by two or more pure tones whose sounds were proceeding in the same direction. We made a careful note of the fact that this wave form represented the displacements of successive particles which occur *at the same instant of time*. A sound wave is always a progressive wave; and we must envisage the associated displacement curve, which represents it, as having a constant shape that moves like a wave on the surface of water. Actually, the particles of air merely swing backwards and forwards in the direction in which the sound travels as the wave passes them.

We have now to consider quite a different kind of wave motion: that produced when the sound of a pure tone travelling in one direction meets sound of equal intensity caused by another pure tone of the same pitch which is travelling in the opposite direction. In each case the motion of the air which would be produced by each tone acting separately can be represented by an associated displacement curve which will be a sine curve. Sine curves representing such sounds are drawn in Fig. 28. One curve, drawn with a thin line, represents a sound travelling from left to right. The other, drawn with a dotted line, represents a sound travelling from right to left. This is evident from their successive positions in the diagrams numbered (i) to (vi) which are drawn to represent the conditions after successive equal intervals of time, each, in fact, equal to

$\dfrac{1}{16}$ the time of a complete vibration of either of the pure tones.

As we are going to combine the curves we must be consistent in our convention, and agree that upward displacements in each curve shall represent actual displacement of particles of air to the right, although the sounds are travelling in opposite directions.

In the first position, marked (i), these curves are drawn at the

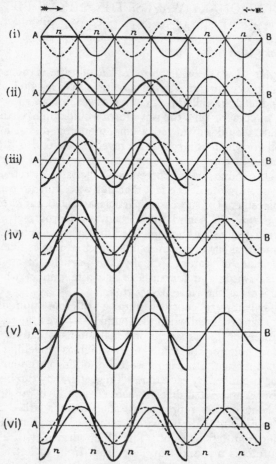

Fig. 28. Diagram of a stationary wave in air (associated displacement curves).

moment when the points of no displacement in each of them coincide. This happens in two positions of the curves, namely, those marked (i) and (v) respectively. We will choose, to begin with, the one in which the displacements in the two curves are in opposite directions. It is obvious that, since the amplitudes and wave-lengths of the two curves are equal in amount, the dis-

placements of the two curves which concern the same particle
of air are not only opposite in direction but equal in amount.
When we combine the two curves we see that there is consequently no actual displacement of any particle of air. All the
particles are in their equilibrium position, that is, in the position
they would occupy if undisturbed by any sound. But it must not
be assumed that they are at rest. Actually, as we shall shortly
see, they would all be moving more rapidly than at any other
time: they would all be swinging at the same instant with maximum velocity through their equilibrium positions.

Now consider the position of the curves when they had both
moved a little in opposite directions as shown in (ii). One curve
will have moved one-sixteenth of a wave-length to the right, the
other one sixteenth of a wave-length to the left. This will happen
very quickly. If the pitch of the tones were middle C, with a
frequency of 256, the time taken to move from the positions in

(i) to those in (ii) would be $\dfrac{1}{16 \times 256}$ of a second, or 0·00024

seconds. If the displacements of the two component curves
are combined, we find that the total displacement of the particles
of air will be shown by the thick-line curve in (ii). After another
equal interval of time the total displacement of the particles of
air will be shown by the thick-line curve in (iii), and after subsequent equal intervals their total displacements will be shown
by the thick-line curve in (iv), (v), and (vi). In (v) it is not possible to distinguish the dotted line for it coincides with the thin
one. In (vi) the thin and the dotted lines have exchanged the
positions they occupied in (iv). Obviously the displacements of
each individual particle of air attain a maximum, at the same
instant, in (v). Thereafter they diminish, and if we were to
draw them for intervals of time to be numbered (vii), (viii), and
(ix), the combined displacement curve would pass successively
through the positions represented in (iii) and (ii) and then return
at (ix) to the position shown in (i). If the student has any difficulty in seeing, with his mind's eye, the return swing through
these positions in intervals of time which would be numbered
(vii), (viii), and (ix), he should actually draw them and so become
quite clear on the point.

The next thing that would happen would be that every
particle which had thus been represented as having swung upwards and back again, would make an equal excursion, to be
represented downwards and back again, in an equal period of

time. Actually, all these particles will have made excursions, on a greatly diminished scale, to the right and back again. They will now make excursions to the left and back again. Similarly, every particle which had been represented as having swung downwards and back again will now be represented as swinging upwards and back again. Actually, all these particles will first have swung to the left and back again; and they now swing to the right and back again.

It will be observed that there is a series of particles, marked *n*, at distances separated by half a wave-length of the component tones, where there is never any displacement at all. The associated displacement curve, representing the combination of the two component curves, bends first to one side of *AB*, then to the other side, as though pivoted at these points. These points are called NODES. Since the displacement curve never moves to the right or left, the wave it represents is called a STATIONARY WAVE.

There is a series of particles half-way between successive nodes which swing backwards and forwards farther than any other particles. These are called ANTINODES.

Starting from an antinode and coming in turn to particles nearer and nearer to a node, we meet particles whose swings are successively smaller and smaller, till at the nodes there are no swings at all. Now consider any one combined displacement curve, say that represented by the thick line in (iii), and examine that part of it which lies between two successive antinodes and passes through a node half-way. We will choose antinodes between which the curve slopes downwards from left to right. Starting with the left-hand antinode we have a series of particles between it and the node all of which have swung to the right. Now start with the right-hand antinode and examine displacements between it and the same node. They are all downwards, and therefore represent particles which have all swung to the left. Thus on either side of the node we have particles which are crowded towards the node. The node is therefore a point of condensation. At the node itself the slope downwards from left to right is steeper than anywhere else. A little thought will show that this must represent a greater condensation than any found elsewhere between our two antinodes. If the student has any difficulty in thinking of a slope as a measure of the degree of condensation, he may find it a good plan to come back to this point after he has reached and studied Fig. 29, which exhibits, on a very exaggerated scale, the *actual* movement of the particles of air.

Now, fixing our attention on the same antinodes and the same node, suppose that the thick-line curve has swung at every point to the other side of AB and to an equal distance. Starting from the left-hand antinode we have a piece of the curve which is everywhere sloping upwards till it comes to the right-hand antinode, and the slope is greatest at the node. All the particles between the left-hand antinode and the node, represented as displaced downwards, will, in fact, have swung to the left, away from the node. All the particles between the right-hand antinode and the node are represented as displaced upwards and have therefore swung to the right, again away from the node. This represents a greater rarefaction than any found elsewhere between the two antinodes. A downward slope from left to right represents a condensation. An upward slope from left to right represents a rarefaction. At an antinode there is no slope. The curve there is for an infinitely short distance always horizontal. At an antinode there is therefore never any condensation or rarefaction. At the nodes, on the other hand, except when the particles are in their equilibrium position, the slope of the curve either upwards or downwards *at any given moment of time*— represented by one of the diagrams numbered (i) to (vi) which we have drawn, and others to be numbered (vii) to (xvii) which the student can now draw if he wishes—is always greater than anywhere else. This means that the nodes are always the points of maximum condensation and rarefaction. At a given node the condensation, starting at zero, increases till it reaches a maximum and then diminishes again to zero. Thereafter there is a gradually increasing rarefaction at the node up to a maximum, after which the rarefaction diminishes again till the zero position is reached, at which there is neither condensation nor rarefaction.

Having thus ascertained exactly the displacements which the different particles undergo, and the way the pressure of the air varies for different particles at different times, we may complete the story by considering the velocity of the motions of the different particles at different times. Examine the shape, as it varies with time, of the piece of the thick line curves in Fig. 31 which lies between two nodes. The whole curve is swinging, and as it were bending out, from its equilibrium position in (i) to its maximum displacement at (v). Obviously it moves farthest between (i) and (ii), a less distance between (ii) and (iii), still less between (iii) and (iv), and a very small distance between (iv) and (v). It is therefore moving fastest when it passes through the equilibrium position in (i), and then it gradually moves more and more slowly till it comes to rest at (v), turns back, and moves with increasing rapidity to the equilibrium position. These are the characteristic features of harmonic motion. They reproduce the appearance of the coupling-rod on the railway engine we mentioned in Chapter VI. If, therefore, the stationary wave could excite progressive sound waves which would reach the ear, we should hear a pure tone, and the associated displacement curve of the progressive wave would have a sine form. We shall see that an organ pipe contains stationary waves which excite progressive sound waves outside it.

The velocity at any given time is always greatest at an antinode: it is always zero at a node.

The motions of the particles of air disturbed by a stationary wave have been worked out in the fullest detail, because a clear apprehension of their nature will remove any difficulty the student might otherwise have in understanding what happens in an organ pipe or a wind instrument. Time spent in drawing the curves on squared paper will never be wasted for that reason; and the practical knowledge the student will gain will be worth many times the description given in this chapter. To make this knowledge quite secure the student may work out, from the associated displacement curves, the actual displacement of the particles lying between three successive nodes. To do this he should increase considerably the distances between the nodes as compared with those in Fig. 28. He should draw sixteen horizontal lines, to represent the lines in which the particles will lie after sixteen equal intervals of time, making up between them a total interval of time equal to the period of a complete vibration: seventeen lines in all. He should mark on each line the equilibrium positions of 21 particles of which the first, the eleventh, and the twenty-first would be at the three nodes. He will find it convenient to make successive equilibrium positions a centimetre apart on his squared paper. On each line he should set off to right or left of the equilibrium positions, measured by the lines of his paper, the actual displacements taken from the associated displacement curves shown by thick lines in Fig. 28. The final result will be a diagram such as that shown in Fig. 29. Careful study of this diagram will confirm all the statements made in the preceding paragraphs of this chapter.

In Fig. 29 curves have been drawn through all the positions of the same particle at different times. These curves are to be distinguished from those drawn in Chapter VI. They are actual displacement curves not associated curves, though, of course, they are very greatly exaggerated; moreover, they represent the displacement of the same particle at different times not the displacement of different particles at the same time. But they will obviously be sine curves, for we have already seen that the motion of the same particle in (i), (ii), (iii), (iv), and (v) of

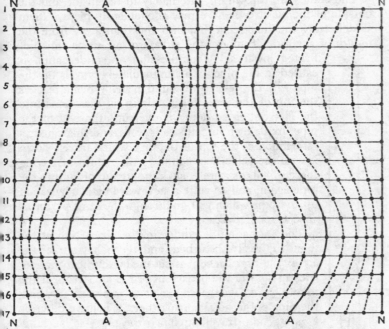

FIG. 29. Diagram of a stationary wave in air (actual displacements very much exaggerated). N=Node, A=Antinode.

Fig. 28 is like that of the coupling-rod of a railway engine seen end-on, in other words, it is harmonic motion; and harmonic motion is represented by a sine curve.

For reasons which are explained in Appendix VII the student is strongly urged to aim at understanding thoroughly the nature of a stationary wave in air, and particularly the behaviour of the condensations and rarefactions in it, before proceeding further.

The Open Pipe of the Organ.

The design of an open flue pipe of an organ is shown diagrammatically in Fig. 30. Air from the bellows of the organ enters the pipe at *a*. It is projected through a narrow slit, *b*, on to a sharp edge, *c*, in what is called the mouth of the pipe. The stream of air in striking this edge, or lip, sets up a series of eddies. Upon the facility with which this series of eddies can be produced depends the quickness or otherwise with which the pipe speaks. It is part of the art of the organ builder so to design pipes that they will speak promptly.

The eddies are thrown off, alternately to the inside and the outside of the air stream. We may conveniently picture them as deflecting the air stream first into and then away from the mouth of the pipe. Suppose the stream of air to be thus deflected inwards. A compression or condensation is caused in the pipe. This immediately travels up the pipe with the velocity of sound in air. When it reaches the top of the pipe it expands in all directions into the air; but in so doing it overruns the end of the pipe, by its own impetus as it were, and this causes a rarefaction at the top of the pipe which travels down it with the velocity of sound to the mouth. Arriving there it will, as it were, suck the stream of air into the pipe again, and so start a new wave of condensation up the pipe which will go through the same performance as its predecessor.

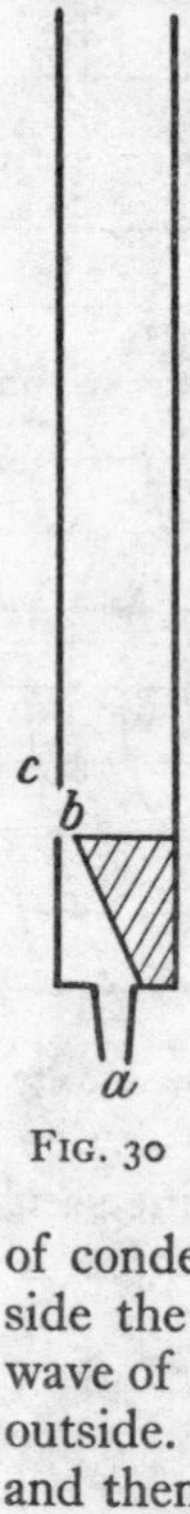

The formation of eddies in the air stream at the mouth is easily disturbed, and the behaviour of the air in the pipe, once it is set in this cyclical motion, tries to control the behaviour of the air stream at the mouth. We have therefore a coupled system in which the pipe is the dominant partner. When we describe a rarefaction, travelling down the pipe, as 'sucking' the air stream into the mouth, it is really insisting on the arrival then of an eddy that is swinging inwards.

FIG. 30

When an eddy is swinging inwards it causes a wave of condensation inside the pipe but a wave of rarefaction outside the mouth. When one is swinging outwards it causes a wave of rarefaction inside the pipe but a wave of condensation outside. The condensations will each increase to a maximum and then diminish to zero. They then change into rarefactions which increase to a maximum and then diminish to zero. To simplify the description we will concentrate on the stages of maximum condensation and rarefaction and call them the condensation and the rarefaction. Half-way between the two eddies in the air stream which cause condensations in the pipe, one is thrown off to the outside and causes a condensation outside the mouth and a rarefaction inside the mouth. As a result a rarefaction is sent up the pipe half-way between the two condensations. Exactly the converse happens to the air outside the mouth. The vibration of the air outside the mouth sets up progressive sound waves which travel to the ear. The vibrations repeat after an interval equal to the time each wave of condensation or rarefaction, travelling with the velocity of sound in air, takes to get to the top of the pipe and back. Thus if the pipe is doubled in length, the period of vibration will be doubled and the pitch of the note heard by the ear will be lowered an octave.

We noted that the eddy stream at the mouth of the pipe would send a rarefaction travelling up the pipe half-way between two condensations. This would begin its journey just as the rarefaction caused at the top of the pipe by the arrival there of the first condensation starts on its journey down the pipe. The two rarefactions will meet in the middle and produce a rarefaction there equal to their sum. On reaching the top of the pipe the rarefaction which is travelling up the pipe will be reflected there as a condensation after slightly overrunning the end. This condensation will start down the pipe at the same time as the second condensation from the mouth starts up it. The two condensations will meet in the middle and produce a condensation there equal to their sum.

At the top of the pipe the air in the pipe is free to move in and out. Consequently the pressure there, or more accurately at a point just outside, will always be the atmospheric pressure. The same thing is true of the mouth. The condensations we envisaged as reaching the top of the pipe and the mouth are always compensated just outside the top and the mouth by rarefactions which are then travelling in the opposite direction. The two cancel out, as it were, and places just outside the mouth and the top of the pipe are therefore places at which particles of air are swinging vigorously towards and away from the pipe, but always at atmospheric pressure.

This makes clear the nature of the motion of the air in the pipe. Two places just outside the mouth and the top of the pipe are places where there is never any condensation or rarefaction, but these places oscillate. These are the characteristics of an antinode in a stationary wave. If we assume that the condensations or rarefactions which meet, as we have seen, half-way up the pipe are for all practical purposes equal in amount, there will be no displacement at this point, but the total condensations or rarefactions there will reach a maximum value. These are the characteristics of a node in a stationary wave. For all practical purposes, therefore, the vibration of the air in the pipe will be that caused by a stationary wave. The behaviour of the particles of air in the pipe at successive intervals of time will be like that represented in Fig. 29 for the nine particles which lie between the two antinodes. The associated displacement curves for the stationary wave in the pipe will be like those drawn between two antinodes in Fig. 28. For the simple oscillation described, these associated curves are sine curves. The progressive waves

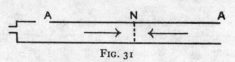

FIG. 31

set up at the mouth by the disturbances there will therefore also be represented by associated curves which are sine curves. The ear will hear a pure tone whose pitch will be determined by a wave-length double the length of the pipe.

It has been noted that when a condensation or rarefaction reaches the top or the mouth, some of it is dispersed into the surrounding air as a progressive sound wave. Energy is thus constantly disappearing from the pipe and is replaced by the air stream which is blown into it. Consequently the pipe ceases to speak very soon after the air stream ceases to be blown into it.[1]

The position of the stationary wave in an open pipe, such as we have described, may be exhibited graphically by the diagram in Fig. 31; the arrows showing the directions in which the particles of air are displaced, as pictured in Fig. 29, during that half vibration when there is a condensation at the node. During the other half vibration, when there is a rarefaction at the node, the directions of the arrows would be reversed. But other stationary waves can be formed in the pipe, always subject to one condition: the mouth and the top of the pipe, or rather points

[1] Since there is constant loss of energy at the open end of the pipe, the condensations and rarefactions which are transmitted down the pipe, by reflection there, must be slightly smaller than the condensations and rarefactions transmitted upwards from the mouth. We assumed them to be equal in the preceding paragraph. This is true as a first approximation; but there must, in fact, be a small motion of the air transmitted up the pipe which is not absorbed into the stationary wave. For the simple oscillation described above, it will be a simple vibration whose wave-length is the same as that of the stationary wave. The loss of energy at the mouth and the open end of the pipe, together with frictional losses of the air in the pipe which produce heat, constitute the *damping* of the pipe.

just outside them, will always be antinodes. Thus stationary waves could be formed in the pipe of which the first two are represented below in Fig. 32; the arrows showing, as before, the directions in which the particles of air are displaced during that half vibration when there is a condensation at the first node. During the other half vibration their directions would be reversed. It will be obvious that the wave formed in (i) of Fig. 32 is that corresponding to a note an octave above the fundamental; that formed in (ii) to a note a twelfth above the fundamental.

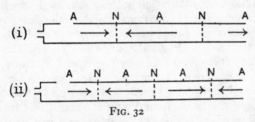

FIG. 32

Thus an open organ pipe can sound all the partial tones of its fundamental. How strong the upper partial tones will be will depend on two things, the width of the pipe and the pressure of the wind. We will consider these in turn.

We described the condensations, propelled up the pipe by the inward deflection of the eddies at the mouth, as overrunning the end of the pipe before they produced a rarefaction which travelled down the pipe. The reflection, as it were, takes place at a point slightly beyond the end of the pipe. As a measure of half the wave-length of its fundamental note the length of the pipe requires a small addition for this reason. This is called the *correction for open end*. Its amount is approximately 0·3 of the diameter of the pipe. This is only an approximation, for the amount varies also with the wave-length. A similar but larger 'correction for open end' is required at the mouth. Two factors therefore affect the correction for open end, the width of the pipe and the pitch of the note sounded. If the pipe were to be blown, in the way explained below, so as to sound its second partial tone as its lowest note instead of the fundamental or first partial tone, the note heard would be slightly sharper than the octave of the fundamental. The air in the pipe would then be vibrating in its octave mode. The free vibration of the pipe in its octave mode would be rather more rapid than that of the first harmonic overtone of the pipe heard when the fundamental is sounded, the difference being due to the alteration in the correction for open end due to the rise of pitch. The divergence would be greater with a wide pipe than with a narrow one.

Now when the fundamental tone is sounded it is the dominating vibration of the pipe. Any attempts made by the air in the pipe to vibrate with the frequency of the free overtones are over-ridden by the influence of the fundamental. The overtones are thereby compelled to conform to the harmonic series. A free overtone, which would be produced if all lower tones were eliminated, would not be so controlled: for reasons explained above in describing vibration in the octave mode, such overtones would become sharper and sharper than the notes of the harmonic series of the fundamental as we work up the series. The vibration of the fundamental, when sounded with them, does its best to stop this. The first overtone would try to be a little too sharp, but as it would then be getting out of step with the fundamental it would be made to keep step by the dominating influence of the fundamental with the result that its amplitude would be reduced. The reason was given in Chapter VIII in describing forced vibrations. The amplitudes of higher over-tones would be still more reduced. Thus with a wide pipe, as the fundamental would allow only overtones in the harmonic series, the first overtone would be weakened, the second weakened still more, while the upper overtones would be checked almost out of existence. A wide open organ pipe has therefore weak partial tones, and only the lower ones are audible. This affects the quality of the note, which is round and dull, lacking brilliance. A narrow organ pipe is less affected in this way, because the 'correction for open end' is so much less; its upper partial tones are stronger and extend higher up the harmonic series. Its tone is richer or more brilliant but has less body than that of a wide pipe.

Organ builders make use of this effect. For stops such as the gamba, for which a stringy tone is required, they use narrow pipes. For a loud and solid tone, which can be made brilliant by adding other stops such as a principal and mixtures to supply deficiency of harmonic overtones, they use an open diapason with a wide pipe.

The quality of the note of a pipe is also affected by the pressure of the wind. Every one knows that if a tin whistle is blown lightly it gives a fairly soft tone. If it is blown more strongly the tone becomes shriller. The pitch also tends to become rather higher when the whistle is blown strongly. If blown still more strongly the note leaps up an octave, and this fact is used by the player to produce notes in the higher octave as required. These changes are due to the changing strength of the overtones. When the whistle is overblown the demand of the first overtone to take charge of the vibrations becomes irresistible and the fundamental, unable as it were to compete any longer, suddenly gives out and disappears. By still stronger overblowing it is possible to cause the second partial tone, or first overtone, to disappear.

Very similar, though fortunately less crude, effects are produced by altering the blowing of organ pipes. Some alteration of quality can be produced by increasing the wind pressure supplied to the pipe.

The author is indebted to his colleague Dr. Kaye, of the National Physical Laboratory, for the following scientific note on overblowing of organ pipes:

'When the blowing pressure of a pipe is increased the facts appear to lie between the two extremes of a gradual increase in the intensity of the octave and a decrease in that of the fundamental, and a sudden change from fundamental to octave with no previous increase in the intensity of the octave.

'The simple theory assumes that the wind stream in striking the lip of the pipe produces a series of eddies. In the absence of the pipe the frequency of production of eddies would increase with the velocity of the stream. In the coupled system of pipe and wind stream, however, the pipe largely controls the frequency and only a slight increase of frequency occurs at first as the blowing pressure is increased. When the velocity of the stream reaches such a value that, in the absence of the pipe, the frequency of the eddies would be nearer to the octave of the pipe than to the fundamental, a sudden jump occurs to the octave frequency. The frequency of the eddies is now controlled by the pipe vibrating in its octave mode.

'In practice, however, the behaviour is more complex (see, for instance, Lough, *Phil. Mag.* xliii, 72, 1922, and Bhargava and Ghosh, *Phys. Rev.* xx, 452, 1922). The intensity of the octave gradually increases as the blowing pressure increases. The intensity of the fundamental at first increases and then gradually decreases for a time, but finally disappears suddenly, leaving the octave alone. Just before the disappearance of the fundamental, beats are heard in the octave.'

The observation in this note about beats heard just before the fundamental disappears is interesting. The beats obviously represent the battle for mastery between the fundamental and the octave which is trying to vibrate in its own mode, slightly sharper than the fundamental can permit. They are the last expiring effort of the fundamental before it dies.

In Figs. 31 and 32 the nodes and antinodes of the fundamental and first two overtones were represented. An antinode is a point at which the pressure is always atmospheric. If in the pipe shown diagrammatically in Fig. 31 a hole were bored in the side at the point where there is a node inside the pipe, the pressure there inside the pipe would become atmospheric pressure; the node would have to disappear and be replaced by an antinode. The vibration of the air in the pipe would then be that shown in (i) of Fig. 32. Consequently the pitch would be raised by an octave. This device is used by organ builders to obtain a particular quality of tone. A suitable 8-foot pipe so treated would be labelled on the draw stop '4 ft.'.

In Chapter VI we calculated that the wave-length of a note whose pitch was that of middle C would be just over 4 feet. If an organ pipe were sounding this note, twice its length plus the corrections for open end at the top and the mouth would equal this wave-length. Consequently the length of the open flue pipe required to produce this note will be about 2 feet. The bottom note of the manual of the organ is two octaves lower. An open flue pipe about 8 feet long will therefore be required to sound it. This is why the stops of the organ which sound notes at normal pitch are called 8-foot stops. Stops sounding an octave higher are called 4-foot stops, those sounding an octave lower are called 16-foot stops. The reason will be obvious.

The behaviour of the air in an open organ pipe has been explored experimentally. At the beginning of the nineteenth century Savart lowered a small tambourine, carrying sand, down an organ pipe. The sand danced about unless it was at a node. It danced vigorously at the end of the pipe which is always an antinode. Later König devised an appliance to demonstrate, by the flickering caused in a gas-flame, the variations of pressure at a node. The experiments are described in books on acoustics.

The Stopped Pipe of the Organ.

Stopped pipes in an organ have a plug at the top end, which can be pushed in or pulled out to tune the pipe. The behaviour of the air in a stopped pipe is different from that of the air in an open pipe. As in the case of the open pipe, we may assume that displacement of the eddy motion at the mouth causes a condensation to travel up the pipe. When it reaches the closed end it is there reflected, without appreciable loss, as a *condensation*. This travels down the pipe to the mouth where it deflects the wind stream outwards, causing a rarefaction just outside the mouth. This rarefaction travels up the pipe, is reflected at the closed end, travels down the pipe, and 'sucks' in the stream at the mouth causing another condensation. This, in turn, travels up the pipe and goes through the same cycle.

The complete cycle thus requires two journeys up the pipe and two down it. Thus to sound a note of the same pitch, a closed pipe needs to be only about half the length of an open pipe. The wave-length of the vibration is rather more than four times the length of the pipe, for to the distance between the lip and the closed end there has to be added the correction for open end at the mouth.

If the end of an open pipe is covered so as to make it into a closed pipe the pitch does not fall quite an octave. This can be explained by the correction for open end in the two cases. Suppose l to be the length of the open pipe from the lip to the top margin; c_1 to be the correction for open end at the top; and c_2 that at the mouth. The wave-length of the fundamental note of the open pipe is $2 (l+c_1+c_2)$. The wave-length of the note an octave lower is $4 (l+c_1+c_2)$. When the top is covered the correction for open end at the top is not required, but that at the mouth remains the same. The wave-length of the fundamental note is $4 (l+c_2)$, which is less than that of the octave below the fundamental note of the open pipe by $4 c_1$ or about $\frac{4}{3}$ the diameter of the pipe. The pitch of the fundamental note of the pipe when stopped is therefore higher than that of the octave of its fundamental note when open, and the wider the pipe the more marked the difference.

Like the open pipe, the stopped pipe can sound harmonic overtones. The correction for open end at the mouth of a wide stopped pipe is greater than that at the mouth of a narrow one. Consequently, for reasons given in the case of the open pipe, the harmonic overtones of a wide stopped pipe are weaker than those of a narrow one. But the series of harmonic overtones is not the same in a stopped pipe. The mouth of a stopped pipe must be an antinode, as it is open to the air. More accurately, a point just outside, which oscillates towards and away from the mouth, is an antinode. The closed end, being a quarter of a wave-length from the mouth, is a node for the fundamental note. There can be no oscillation up and down the pipe there, for oscillation is prevented by the plug. It is evident that, as there is oscillation up the pipe when a condensation is travelling up it, the particles of air take up successively the positions which are shown in the top quarter of the diagram in Fig. 29 between the left-hand antinode there shown and the middle node. When the condensation begins to travel down the pipe, the particles of air take up, successively, the positions shown in the second quarter from the top of this diagram between the same antinode and the middle node. Thus the closed end is a place of maximum condensation. In the second half of the vibration it becomes a place of maximum rarefaction.

The position of the stationary wave in the closed pipe may be exhibited graphically by the following diagram, the arrow indicating, as before, the direction in which the particles of air are displaced during one half-vibration.

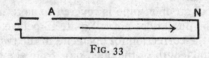

FIG. 33

But other stationary waves can be formed in the pipe, always subject, as we have seen, to two conditions: the mouth will always be an antinode and the closed end a node. Only a condensation or a rarefaction can be reflected at the closed end in the way with which we became acquainted in Chapter VII; and at an antinode there is never any condensation or rarefaction. The effect can be studied from Fig. 29 by supposing the vertical line there drawn through the right-hand node to be the surface of the closed end of the pipe. Stationary waves can thus be formed in the pipe of which we can show the first two in Fig. 34 below, the arrows indicating, as before, the directions in which the particles of air are displaced when there is a condensation at the first node. During the half-vibration when there is a rarefaction there the directions of the arrows would be reversed.

It will be obvious that the wave formed in (i) of Fig. 34 has a wave-length equal to one-third of that of the fundamental, that is, it corresponds to a note a twelfth above the fundamental tone of the pipe. Similarly, that formed in (ii) corresponds to the seventeenth of the fundamental. These are the third and the fifth tones of the harmonic series starting from the fundamental. The tones corresponding to the even numbers of the harmonic series are lacking. This gives to the stopped pipes, whether wide or narrow, their characteristic qualities. If the pipe is wide, as in the stopped diapason, the first overtone is weak, and the second is almost inaudible. Such a pipe sounds an almost pure tone. When a stop consisting of such pipes is drawn with a stop consisting of open pipes of narrow bore, such as a dulciana, it gives the pleasant creamy tone familiar to organists; since the dulciana supplies the deficient overtones, while the strength of the fundamental in the combination prevents any thinness of tone.

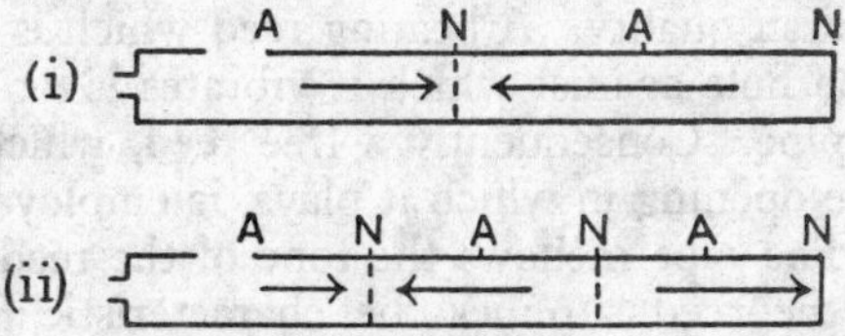

FIG. 34

These considerations indicate the general principles of the various flue pipes of the organ, for the details of which reference may be made to books on that instrument.

The Reed Stops of the Organ.

In a reed stop on the organ, a reed fixed in a small chamber at the foot of the pipe beats against the opening from the chamber into the pipe. The effective length of the reed can be adjusted by a wire the end of which projects from the top of the chamber. This is used to tune the pipe. By its side the pipe, usually of a conical shape expanding upwards, emerges from the chamber.

The behaviour of the stationary wave in the pipe is similar to that in a stopped flue pipe sounding its fundamental note. When the pressing down of the key in the manual admits wind to the chamber, some of it enters the pipe carrying with it the reed which then closes the mouth of the pipe: a condensation is thus started up the pipe. This travels up the pipe and is reflected from the open top as a rarefaction. This rarefaction travelling down the pipe keeps the reed against the mouth when it gets there, and is reflected as a rarefaction. Reaching the top again it is reflected at the open end as a condensation; and this condensation, travelling down the pipe again, releases the reed, which thus makes its own vibration. The wave-length of the note is four times the length of the pipe.

Neither the pipe nor the reed is in sole control of the vibrations in the pipe. The reed is fairly stiff and there is mutual constraint between its motion and that of the air in the pipe. If the tone is to be satisfactory and the pipe to speak readily the reed and the pipe should have vibrations of approximately the same rate.

A reed stop is rich in the upper partial tones which give to it its characteristic quality. The reed at the mouth, combined with the air-pressure in the chamber of the pipe, makes the pipe equivalent in length to a stopped pipe; but, as Helmholtz showed, a conical pipe, unlike a stopped cylindrical pipe, has the full complement of harmonic overtones. The reed itself has a note of harsh quality. A beating reed which is large enough to cover the hole against which it vibrates is too harsh for use without a pipe. Consequently a free reed, which can vibrate through the opening in which it plays, is employed in the harmonium. The pipe mellows the tone of the reed in the organ pipe; but the reed supplies the characteristic higher partial tones to the vibration of the air in the pipe.

Orchestral Wind Instruments.

The principles which govern the quality of the wind instruments of the orchestra will be evident from those of the various organ pipes.

The flute is in principle very similar to the open pipe of the organ. The air stream is supplied from the lips of the player which direct it over the hole in the mouthpiece. The D flute of the orchestra is lengthened to carry two extra keys which provide two notes, C and C♯. The principle of the instrument is the same as that of the familiar tin whistle which, of course, has no such keys. We will therefore describe the behaviour of the tin whistle with its six finger holes, as it is simpler to explain acoustically. When the finger holes are all closed the air between the lip of the opening in the mouthpiece and the open end sounds the fundamental note of the whistle. The top finger hole lies half-way between the lip of the opening in the mouthpiece and the open end. When it is uncovered an antinode is formed in the middle of the pipe which vibrates in the octave mode. A similar effect was noted above in one class of open flue pipes in the organ. Other notes of the lowest octave are produced by uncovering the finger holes one by one, beginning at the bottom. This reduces the speaking length of the pipe. By overblowing, the notes of the next octave are produced. The effect of size and position of the holes of different designs of flute are discussed in works on orchestral instruments.[1]

In Chapter VI we drew diagrams to show, in two different cases, the combined associated displacement curve for a note consisting of a fundamental tone and its first harmonic overtone.

[1] See Appendix VIII.

In the second, Fig. 25c (v), the vibration of overtone had only one-fifth of the amplitude of that of the fundamental. In the first, Fig. 25b (iii), it had two-thirds the amplitude; and there was combined, with the result, the vibration of the second harmonic overtone with half the amplitude. It was stated that the final curve, obtained graphically, in Fig. 25c (v) was like the trace, obtained by Miller's phonodeik, from a flute sounded piano; while the curve in Fig. 25b (iii) was very like the trace of a flute blown forte. The two wave forms may be examined in con-nexion with the note, earlier in this chapter, on overblowing an organ pipe. They throw an interesting light on the note, which, in turn, explains the difference between the two wave forms.[1]

Reed instruments used in an orchestra, such as the oboe and the clarinet, are in general principle not unlike the reed pipe of the organ; but the reed is of necessity very flexible since it has to respond to all the notes of the instrument, and its vibrations are controlled by the stationary waves in the tube. The oboe and the bassoon have conical tubes, and so possess a complete series of harmonic overtones. If they are overblown, the air in the tube vibrates in the octave mode. The clarinet has a cylindrical tube; and, like a stopped organ pipe, it therefore sounds only the odd partial tones. This gives its note the quality which distinguishes it from instruments like the oboe. When the clarinet is overblown its note leaps from the fundamental to the twelfth. But this is not quite the whole story. The statement that only odd partial tones are sounded is quite true of the primary overtones of the clarinet, which determine the effect of overblowing. But the bell at the open end of the tube, and a slight taper near the mouth-piece, enable the reed to set up very small secondary vibrations in the tube as forced vibrations. These supply weak even partials, which are therefore to be heard very faintly in the note of the instrument. Their vibrations are much too weak to have any say in the effect of overblowing.

In the brass instruments such as the horn, trombone, and trumpet, the lips of the player are set in vibratory motion and take the place of the reed in wind instruments. A better analogy would be the production of the voice by the vocal chords, since in both cases membranous structures take part in the vibrations set up: the only examples of the musical use of membranes whose vibrations are actuated by a wind stream. Brass instru-ments depend for their notes on the sounding of the harmonics of their fundamental, with accompanying upper partials of the

[1] See p. 117 and Appendix V.

note sounded. When the bore is narrow compared with its length, the fundamental is difficult to blow, but the upper partials are easily sounded. It is the necessity for length which requires the bending of the tube referred to in Chapter VII. It would be outside the scope of this book to attempt any further description of the brass instruments.

The nature of the analogy between the effect of bending in speaking-tubes and that of bending in tubes of wind instruments is now evident. In both cases particles of air oscillate backwards and forwards along the tube. In a speaking-tube the oscillations are transmitted along the tube from particle to particle in a progressive wave. In a wind instrument the oscillations reach a maximum value at an antinode and there is no oscillation at a node. The bending of the tube has the same effect on the oscillations in each case. The motion of the air in a post-horn is like that shown in Fig. 29. The oscillations take place along a straight line. In the orchestral horn the oscillations in a bend are deflected by the tube just as they are in a speaking-tube. The stationary wave in a wind instrument is therefore bent by the tube when the tube bends. When the tube straightens out the stationary wave straightens out. In both speaking-tube and wind instrument the tube prevents most of the energy in the air from escaping.

The flexibility of intonation in the wind instruments of the orchestra is important, as we saw in Chapter V, because it enables them to adapt themselves, more or less completely, to the musical scale system. A full account of this flexibility would require a description of the technique used by artists who play the various instruments. But what we have now learnt of acoustics will enable us to appreciate the nature of this flexibility and so to understand in a general way the intonation of the orchestra. The intonation of the flute has a marked degree of flexibility, which depends on the position of the lips as well as on the pressure at which the breath is expelled through them. Experiments which the reader can make by blowing hard or softly, across or slightly into, the mouth of a narrow necked bottle will show how the note of the bottle can be made to vary. Something similar happens to the flute, which may be regarded as a form of resonator. So flexible is its note that in the hands of an unskilful player the flute may be played badly out of tune. Like the flute, the reed instruments are keyed instruments and in their lowest register are adapted for equal temperament, the notes of the upper register being produced by overblowing

combined with suitable fingering. The flexibility of intonation
of the oboe and clarinet is limited, but it can be controlled
somewhat by the pressure of the lips on the reed. To that
extent there is some mutual constraint between the vibrations
of the reed and the pipe such as we found to be so marked in the
reed pipe of an organ. The notes of the bassoon, which is a
large oboe, are more adjustable than those of the smaller instrument; and in the upper registers some of them can be altered
slightly by alternative fingerings. In the trombone one part of
the tube slides inside the other. Its intonation is quite free, like
that of the strings. The brass instruments, in general, use their
natural partial tones, as we have observed; and if the tube is
correctly shaped these are given by the harmonic series. Their
notes are sharpened by pressing the lips together in the mouthpiece[1] and flattened by relaxing their pressure: to some extent
a free vibration can be 'forced' by the player. The higher the
frequency of the note the less easy this modification of it
becomes. The effect and tuning of the valves is a study in
itself: it is described in books on these instruments. In the horn
the player's hand, held in the bell, can be used to flatten the
note by closing the tube a little, which we may compare with an
alteration in the 'correction for open end' (see Appendix VIII).
These observations may help to explain why Stanford, in his
Musical Composition, writes: 'To one who has absorbed the
compromise known as "equal temperament" it will come as
something of a shock to find that his idea of the scale is confined
to keyed[2] instruments, and that in the orchestra . . . it has no
place.' And later: 'He may be surprised that chords (e.g. in
Wagner), which sound crude on his piano, lose all their roughness in the orchestra . . .'

[1] A simple account of the mouthpieces of different brass instruments, and
of their effects on intonation, will be found in *Acoustics of Orchestral Instruments,* by E. G. Richardson.
[2] That is, as the context shows, what we have usually called 'keyboard'
instruments.

X

VIBRATION OF STRINGS

IN Chapter I it was stated that a flexible string can vibrate as a whole, as two half lengths, three thirds of a length, and so on; that the vibration of the string as a whole gives its fundamental tone; and that the sectional vibrations give rise to harmonic overtones. Equipped with the information we have now collected, we are in a position to inquire into these statements. To do so will enable us to form an idea of the effect, on the quality of the note heard, of different methods of exciting vibration in a string; as for example by plucking, bowing or striking it with a pianoforte hammer. This of course is the aspect of the vibration of strings which is of interest to the musician.

It may here be observed that, if the string is not perfectly uniform, the overtones will not fall into the harmonic series. If the string is uniform but not very flexible, the overtones will be weaker: the stiffer the string the weaker will the overtones be.[1] Moreover, if we use stiffer and stiffer strings the more will their overtones become sharp on the notes of the harmonic series, and increasingly so as we reach them in turn moving upwards.[2] The use of strings so stiff that this last effect could be perceived is avoided in musical instruments.

A uniform and flexible string set vibrating in any of the ways mentioned in the first paragraph above, gives a sound which is different from the fundamental tone of the string. The sound heard is a complex note containing many of the harmonic overtones, upon the number and relative strength of which depends the quality of the note heard. The complex progressive wave which the vibrating string, assisted by its soundboard, causes the air to transmit to the ear is analysed by the ear into its constituent pendular vibrations or harmonic motions; and the sensation caused by the cumulative effects of all these vibrations gives to the note heard its characteristic quality. It is possible by mathematical analysis to obtain, from the altering shape of a vibrating string, information about the frequencies and the intensities of the various pendular vibrations which will reach the ear as constituents of the compound wave in the air. Such mathematics is quite beyond the scope of this book, but

---

[1] See Appendix X, p. 168.
[2] See *Theory of Sound*, Rayleigh, 2nd ed., §§ 137 and 189.

by the use of squared paper, and by appeal to what we have learnt of the power of the ear to analyse notes, we shall be able to exhibit the nature of the motion of vibrating strings.

For a string so to vibrate that it emits its fundamental note as a pure tone, is a physical possibility. The vibration is the sympathetic one excited in the string by a pure tone whose frequency is the same as that of the string vibrating as a whole. It can readily be produced if a tuning-fork, to which the string has been tuned, is set in vibration and its stem held against the bridge of the soundboard over which the string vibrates. If the tuning-fork is then removed, the string and soundboard will continue to emit the pure tone first excited by the fork. When the string is so vibrating, as a whole, its vibrations can be represented by a diagram as follows:

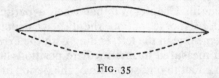

FIG. 35

As usual, the extent of the displacements of the particles of the string by the vibration excited have been greatly exaggerated to render their nature more obvious to the eye. The upper, continuous line represents the extreme upward limit of vibration, the lower dotted line the extreme lower limit. These two lines enclose the fuzzy shape which the eye sees in looking at a vibrating string. These two curves have been drawn as sine curves. We want to produce a pure tone: this suggests that we need the simplest possible vibration; and we have already learnt that the sine curve represents to the mathematician the simplest vibration he can find. It is therefore reasonable to start with a displacement which is a sine curve; but, in fact, the mathematician would tell us that a string released from the position shown, in Fig. 35, as a sine curve would vibrate in such a way as to sound a pure tone.

It was stated in the Introduction that to discover the motion of a vibrating body the mathematician invokes the aid of the science of dynamics. We have now met an example of a problem in vibrations which the mathematician attacks in this way. At the end of this chapter we shall find an indication of one of his lines of attack. Meanwhile let us assume, what the mathematician knows, that a string released from the position shown in Fig. 35 as a sine curve will sound a pure tone; and applying what we have learnt of the ear's power of analysing sound let us see what it can tell us about the way in which the shape of the string must alter as it vibrates, if our assumption is correct, and about the problem which the mathematician is able to solve.

It is evident that since the ear hears only a pure tone the waves reaching it through the air can contain no components other than the harmonic vibration equivalent to the pure tone heard: in other words, the wave in the air must be represented by a pure sine curve. If the vibration of any part of the string were other than a harmonic vibration with a period the same as that of all the other parts, this would not be the case. We therefore reach the conclusion that each particle of the string must be vibrating with harmonic motion, the frequency of the vibration being the same for each particle.

Now we saw in Chapter VI that when a particle of *air* is vibrating with harmonic motion, the displacements, beginning with the extreme one to the left, ending with the extreme one to the right, and measured in intermediate positions at successive intervals each equal to one-sixteenth of the period of a complete vibration, will be as follows:

$$-1, \ -0\cdot92, \ -0\cdot71, \ -0\cdot38, \ 0, \ 0\cdot38, \ 0\cdot71, \ 0\cdot92, \ 1$$

the process then reversing, so that the displacement passes through all these values again in succession till it reaches -1. If all the particles of the *string* are vibrating with the same harmonic motion they must each of them pass through a series of displacements whose ratio to the maximum displacement of the particle is given by our series of decimals; and they will do so after intervals of time equal to one-sixteenth of a complete vibration. Since in Fig. 35 we start with a displacement above the line, which we shall naturally call $+$, the vibration of a particle of the string will be regarded as beginning at $+1$, at the right-hand end of the series of figures given above, swinging through the series to -1, and returning in the period of a complete vibration to $+1$.

We may insert in Fig. 35 the intermediate positions of a number of particles at the end of successive sixteenths of a period, the displacement of any given particle at the end of the third sixteenth of a period, for example, being $0\cdot38$ of its displacement to begin with. A curve drawn through the positions of a number of particles of the string at the end of the first sixteenth of a period is shown with the number 2 against it in Fig. 36, those drawn through the positions at the end of the second and third intervals are shown respectively with the numbers 3 and 4 against them, while the corresponding curve after the fourth interval will be the straight line $N_1 N_2$ itself, numbered 5. The curves numbered 1 to 5 in Fig. 36 therefore represent the shapes of the string at the beginning and end of the first four intervals each equal to one-sixteenth of the period of a complete vibration. The last one is a straight line and since the first is a sine curve all the others must be, for they are drawn in the same way though to a different vertical scale, that is, a scale successively 0·92, 0·71, and 0·38 that of the scale of curve 1. The string will pass in turn through similar shapes below the line $N_1 N_2$, to be numbered 6, 7, and 8, until in 9 it reaches its extreme position below the line. It will then return to 1 through the whole series.

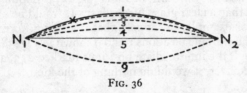

FIG. 36

These curves represent the *actual* displacements of each particle of the string at the same instant of time. We can draw an *actual* displacement curve for a particular particle of the string at successive intervals of time if we take the displacements for that particle as shown by successive curves and then draw vertical lines of equivalent lengths at seventeen equidistant positions on a straight line thus:

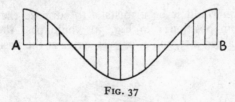

FIG. 37

In this figure we have taken the displacements of the point × in Fig. 36, doubled them to make them more conspicuous, set them off in turn against the seventeen equidistant points in the line AB, and drawn a freehand curve through their ends. We have worked this out in detail, for when we come to more complicated vibrations we shall find it useful to examine the actual displacement curves of particular points in the string.

K

It will now be observed that we have found a vibration in which the string takes, at every instant, a shape which is a sine curve, and that the displacement curves of each individual element of the string are also sine curves. No vibration better calculated to produce a progressive wave representing a single harmonic motion, and therefore producing a pure tone, could be imagined.

Since the vibrations of a string are transverse it is natural to picture them in terms of the simplest vibrations with which we are all familiar, for example those of a pendulum or a swing, and to suppose the kind of vibration we have described to be normal. In fact it is an absolutely special case. If the student is really to understand the difference in quality of the sounds produced by strings vibrating in different ways, nothing is more important than a clear grasp of this fact. It would be a complete mistake, for example, to suppose that if a string were vibrating between the extreme positions 1 and 9 shown in Fig. 38, it would pass through the intermediate positions marked 2, 3, and 4. As we shall see later, it would do nothing of the kind.

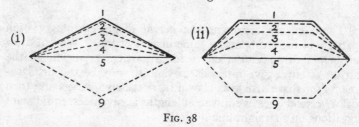

Fig. 38

Still more would it be a mistake to suppose that a string released from position 1 in Fig. 39 would pass through the positions marked 2, 3, and 4, or ever arrive at the position there marked 9.

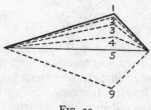

Fig. 39

Here again the ear can help us. If strings, vibrating freely on being released from the positions shown in Figs. 38 and 39, made vibrations like those indicated by the dotted lines, all their elements would be vibrating with harmonic motion and they would all sound their fundamental note as a pure tone. Similar assumptions would have to apply to the free vibrations of strings released from all sorts of shapes, and we should have to suppose that they would all sound their fundamental note as a pure tone. This is contrary to all experience. We may infer that the vibration shown in Fig. 36 is the only one which gives a pure tone for the fundamental note of the string. This in fact is what the laws of dynamics prove.

The problem we must set ourselves to solve is how strings displaced as shown in (i) of Fig. 38 or in Fig. 39 would actually vibrate if left to themselves. We are dealing only with strings which have been excited by a sympathetic vibration, or by plucking or striking and are then allowed to vibrate freely. The problem of the violin string, which is excited continuously to make vibrations is a different one. When we have solved our problem we shall find that we have a perfectly simple rule for finding out, graphically, the free vibrations of a string if we know its initial displacement. We shall be able to construct actual displacement curves for the positions of individual elements in successive equal intervals of time. Hence we shall be able to infer the existence of a long series of harmonic overtones in the note sounded by strings vibrating from the positions shown in Figs. 38 and 39, it being always assumed that the strings are mounted on a soundboard which makes their tones properly audible.

Before we examine vibrations caused by displacements such as those shown in Figs. 38 and 39, let us take a simpler case and see how the problem arises then. Consider what happens to the string when it is vibrating in its octave mode: for example, if it is excited to give sympathetic vibration by a pure tone an octave above the fundamental note of the string. As Pythagoras showed, a string divided into equal parts by a bridge will sound a note an octave above the fundamental note of the whole string. It is natural therefore to find that when a string is vibrating in its octave mode it can be touched exactly at its middle point, for example by a fine camel's-hair brush, without affecting its vibration. If a string is vibrating in its octave mode, the middle point will therefore be a node; and the limiting positions of the vibration will be as shown in Fig. 40, in which we have supposed the vibration to have half the amplitude of the vibration of the string as a whole.

Fig. 40

Between these two positions each half of the string will vibrate with a motion similar to that of the string as a whole, but will do so twice as fast. Thus in the time that it would take the string, vibrating as a whole, to pass from position 1 to position 3, the string vibrating in its octave mode will pass from the equivalent of position 1 to that of position 5, that is, to the equilibrium position represented by the straight line $N_1 N_3 N_2$. *At any moment the vibration of the right-hand half will be in the opposite direction to that of the left-hand half.*

If we combine the initial positions of the vibrations in Figs. 36 and 40, we obtain as the initial position of the combined vibration a curve as shown by the thick line in Fig. 41. What are the next positions of the string?

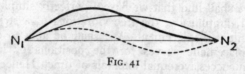

Fig. 41

To understand what happens we must now consider what happens when a wave is sent along a string. With a string vibrating so as to sound an audible note it is not possible to see this motion as it is too rapid. But, as every musician knows, if he thinks of the D and the E strings on a violin, the heavier a string, other things being equal, the lower its note; that is, the more slowly it vibrates. Again, the more a violin string is screwed up, and the greater in consequence its tension, the higher is its note. To examine visually the motion of a wave running along a string we want one which vibrates very slowly, and for this purpose we must choose a heavy string under very little tension. A garden hose will serve our purpose admirably; but any fairly stout piece of rubber tubing some yards long or the spring described on p. 138 will serve.

Take a garden hose fastened at one end and held off the ground by the hand at the other end, preferably with sufficient tension to prevent it sagging too badly. Give a small and rapid shake to the end held in the hand. A wave is seen to travel to the other end of the hose, where it is reflected upside down and travels back to the hand. The student can try the effect of a variety of motions of the hand. He will find that if he sends a double wave consisting of a crest followed by a trough, the returning wave will consist of a trough followed by a crest.

Turn again to Fig. 36, which showed the vibration of a string when sounding a pure tone which was the fundamental of the string, and compare it with the associated displacement curves obtained in Chapter IX by combining two sine curves of the same period and amplitude travelling in opposite directions (Fig. 28). The vibration shown in Fig. 36 is clearly a stationary wave in the string, corresponding to that part of the stationary wave in Fig. 28 which lies between two successive nodes. The octave vibration shown in Fig. 40 is clearly a stationary wave in the string, corresponding to that part of the stationary wave in Fig. 28 which lies between one node and the next node but one, leaving the intervening node in the middle of the string. The scale to which the displacements were drawn in Fig. 28 was much more exaggerated than that used in Fig. 40, and of course we are now dealing with actual transverse displacements of the string, not with an associated displacement curve used to represent, diagrammatically, the longitudinal vibrations in a stationary wave in air.

To explore this further, let us return to our garden hose. Suppose that a crest is formed in it by raising the hand rapidly and with uniform speed, and then reversing the motion instantaneously and dropping the hand with a smaller but uniform speed. A V-shaped crest would be formed, though the angles would be rounded by the stiffness of the hose. If we ignore this rounding effect, the shape of the crest would be as shown in Fig. 42, 1 of which represents it just before it reaches the other end of the hose. Now the other end of the hose is fixed and as the crest arrives there its shape has to be altered, because it can cause no transverse displacement at the fixed end of the string. Such an alteration to its shape would be produced if it then encountered an exactly similar trough travelling in the opposite direction. Such a trough could be imagined as travelling to meet it at exactly the right moment if another length of garden hose had been fastened at the fixed end so as to extend in the opposite direction to our piece of hose, and if a hand, exactly the same distance behind the fixed end as our hand is in front of it, had caused in this 'image' piece of hose a trough to travel to the fixed end at exactly the moment our hand caused a crest to travel there. We could then imagine the fixed end to be unfastened, but (apart from the sagging of the doubled piece of hose which we will ignore) the end which was previously fixed would still remain motionless, for it would be a node when the crest travelling in one direction met the trough travelling in the other. The motions, showing exactly how a crest is thus reflected as a trough, are shown in Fig. 42.

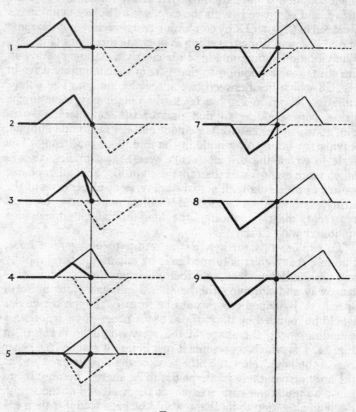

FIG. 42

We are now in a position to find out how a string vibrates, provided we know the initial form of its displacement. We saw two paragraphs ago that a typical vibration could be regarded as a stationary wave with nodes at the ends of the string. We have seen in the last paragraph what form must be taken beyond the ends of the string by the progressive waves which we must suppose to combine to constitute the stationary wave. We have to visualize the wave as extending beyond the ends, the shape of the extensions beyond the ends being reflections of the wave at the ends of the string but inverted. To illustrate this we may consider the initial form of the vibrating string drawn in Fig. 39. Let us draw this with the amplitude reduced by one half and then indicate beyond the ends of the string the waves which we must visualize as the inverted reflections there. The result will be as in Fig. 43, in which the scale is reduced to fit the page.

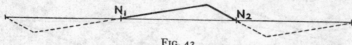

FIG. 43

Now suppose that this curve represents two wave forms of identical shape, one of which is travelling to the left and one to the right, with the same speed, and that at the moment represented in Fig. 43 the two wave forms coincide. Let us then draw, at the end of successive intervals each equal to one-sixteenth of a complete vibration, the positions of these wave forms and combine them. Their next two positions after that shown at 1 in Fig. 44 will appear as shown in 2 and 3.

Proceeding in this way, the remaining combinations at intervals of one-sixteenth of a complete vibration have been drawn in Fig. 44. The result is interesting. Clearly the vibration has no resemblance to the vibration, sketched in, in Fig. 39. It is easy to see the displacement running along the string to both its ends. The displacement in the thick line in 1 of Fig. 44 is seen, as it were, to divide into two. The one half travels to the left-hand end, and any element of the string between the initial displacement and the end remains quite stationary till the displacement, running along the string, reaches it. The same thing happens to the right-hand part of the string. The piece of string behind the two displacements collapses into a straight line. When the right-hand displacement has reached the right-hand end, this straight line runs across the string as it were, reaches the left-hand end, goes through the reverse process and finally produces the outline shown in 9 of Fig. 44. The whole process then returns through the same outlines to the original position, for 1 and 17 are identical.[1]

We may construct the actual displacement curves after successive intervals of time, for each element of the string. The curves, all composed of straight lines, for seven elements

[1] This method of showing how a string vibrates freely was first used (*Phil. Trans.*, 1800) by Thomas Young, whose name has been made familiar to us by the modulus of elasticity (*Theory of Sound*, Rayleigh, 2nd ed., § 146).

one-eighth, one-quarter, three-eighths, half, five-eighths, three-quarters, and seven-eighths of the way along the string from the left-hand end are given in Fig. 45, and there numbered from 1 to

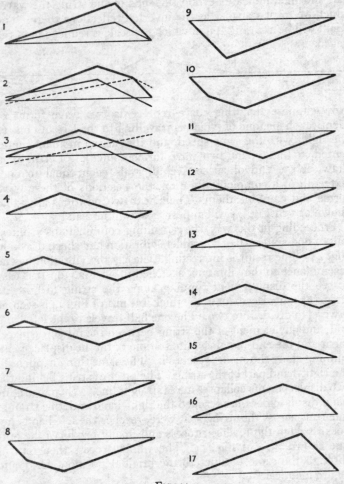

FIG. 44

7. Observe that, in this figure, 7 is the same as 1 turned upside down but moved along half a phase, 6 the same as 2, and 5 the same as 3. To exhibit this we have added, by dotted lines, to 5, 6, and 7 the next half vibration.

The vibrations shown in Figs. 44 and 45 are those of a string

which is plucked one quarter of the way along it from one end. Those in Fig. 44 form a complicated stationary wave, in which

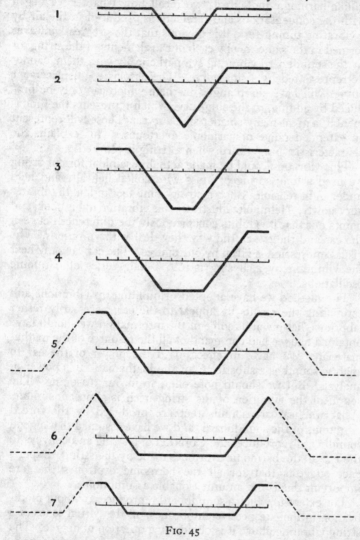

FIG. 45

the ends of the string are nodes. The vibration of the string of a harpsichord is of this nature; but there would be objections to the choice of one quarter of its length for the point at which to pluck the string, as we shall see.

It is evident that the vibrations of the elements are quite unlike harmonic motion. If we recall how Chapter VI began with the condensations and rarefactions produced in the air by a vibrating tuning-fork, it is obvious that the progressive waves formed in the same way by each of the elements of the string, or by the soundboard vibrating sympathetically with them, cannot be represented by a sine curve. The associated displacement curves will have sharp angles in them which can only be produced by adding, to the sine curve which represents the fundamental, a great many smaller sine curves. These will represent an extended range of harmonic overtones. This explains the characteristic 'ping' heard when a string is plucked.

This vibration could be made actually visible if for our string we used a long spring of brass wire coiled helically and held under light tension. Such a string some feet in length vibrates very slowly. Helmholtz illustrates this vibration in his book, and points out that it exhibits conspicuously the difference between the ear and the eye in the way they deal with vibrations. The visual impression is given by our figures. The ear can only hear the vibration by analysing it into a long series of harmonic oscillations.

The method we have adopted of building up vibrations and estimating the results by appeal to the ear, necessarily rather laborious, has given us some of the information we should have obtained had we had command of all the resource of the mathematician. We have, in effect, used the power of the ear to analyse sound as a substitute for the mathematician's power of analysis. But we should note one gap in our reasoning. The sound of the vibration of the string itself is almost inaudible. What we hear are sounds that are produced by the forced vibrations of the soundboard; and we have assumed that a good soundboard reproduces the vibration of a string and conveys to the ear all the harmonic overtones made by the string. Experience suggests that for all the harmonic overtones that are important this is not an unreasonable assumption.

The extent to which a soundboard reproduces, without modification in their respective amplitudes, all the partial tones of a string vibrating above it is obviously a question of degree. The kind, quality, and thickness of the wood to be used in the soundboard of a piano has been determined empirically by makers of the instrument. The soundboard is an important factor in producing good tone; and we have learnt that the difference between a good and a poor tone must be represented physically by differences in the number, the selection, and the amplitudes of the partial tones transmitted, in a compound wave, to the air by the soundboard. The difference between a first-rate violin and an indifferent one, the strings of which may be identical, depends entirely on the efficiency of the instrument as a coupled system; and the examination of the causes of this efficiency is quite beyond the powers of mathematical analysis. Knowledge of the wood to use and the way to shape it, and so forth, constitutes the craftsmanship of the maker which in a master-hand far transcends any knowledge that science can offer.

It is interesting to see how the bridges, on which the string drawn in Fig. 44 is resting, are made to behave. Take the right-hand one. For a quarter of a period it is pulled upwards at a certain angle. Suddenly it is pulled downwards at a smaller angle and is so pulled for three-quarters of a period.[1] Then the process repeats. Yet this simple motion combined with that of the bridge at the other end suffices to excite vibrations in the soundboard which the ear hears as a highly complex note.

We now see how to find the vibrations executed by a string displaced as shown in Fig. 41. We draw the inverted reflections of this curve at each end and put them through the same process as that adopted for the plucked string (see also Appendix IX). The first five positions which the string will take up after intervals of one-sixteenth of a complete vibration are shown in A of Fig. 46. The next four positions will be seen by turning the book upside down and looking in turn at the first four positions in the order (iv), (iii), (ii), (i) as they will then appear, upside down. In the second half vibration the string returns through these positions, in order, to its original position. It may be verified, by trial, that the same thing would be true for Fig. 44. It is interesting to observe that the thick line in (iii) is the same shape as the dotted line marked 3 in Fig. 36 and that the thick line in (v) is the same shape as the dotted line in Fig. 40. The reason will be evident to any one who works out the vibration by the alternative method explained in Appendix IX. In B there are shown the actual displacement curves, after all sixteen intervals, of three elements of the string, namely those one-quarter,

[1] Number 3 exhibits the arrival of the angular displacement at the right-hand end of the string, when the direction of the string at this end changes suddenly from that shown in 2 to that shown in 4. The direction at this end shown in 3 is a stage through which the string passes instantaneously. Similar observations apply to 15, and for the left-hand end to 7 and 11. It is as though an instantaneous photograph had been taken of the string, for 3, 7, 11, and 15 of our diagram, in the middle of these changes.

A

(i)

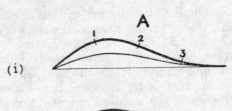

(ii)

(iii)

(iv)

(v)

B

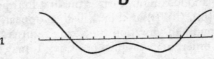

1

2

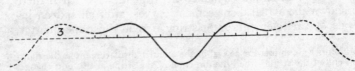

3

FIG. 46

one-half, and three-quarters of the length of the string measured from the left-hand end. The middle one numbered 2 is obviously a sine curve; and the middle element of the string will therefore contribute, to the compound wave which reaches the ear, a pure tone corresponding to the fundamental. The first and the third ones are not sine curves at all. Consequently the motion of the elements of the string one-quarter and three-quarters of its length from either end is not a simple harmonic motion. The wave is very like that obtained in Chapter VI by combining the sine curves of a fundamental and its octave as shown in (vi) of Fig. 25c. These elements of the string will therefore contribute, to the compound wave that reaches the ear, a wave motion of which the constituents are the sine curves for the fundamental tone and the octave tone. The curve of 3 is exactly the same shape as the curve of 1 turned upside down, but it differs by half a phase. This phase difference between corresponding elements in the left-hand and right-hand half of the string is always found when there are even partial tones in the note caused by the vibration. This will emerge in the next paragraph in which we shall find a vibration without this phase difference.

As a test, to satisfy himself that he has grasped all this, let the student similarly find the successive shapes of a vibrating string of which the initial displacement is given by (i) of Fig. 38. He should then draw actual displacement curves showing the displacement of particular elements of the string at the end of successive sixteenths of the period of a complete vibration. He will find that the right-hand half of the string is always moving in the *same* direction as the left-hand half (see Fig. 47). Now this is a very important observation. We have found the successive positions of the string vibrating in its octave mode, as shown in Fig. 40, and we saw that at any moment the vibration of the right-hand half will be in the *opposite* direction to that of the left-hand half. The different behaviour of the displacement curves in Fig. 47 means that when a string is plucked at its middle point, the octave vibration and consequently the second partial tone must be missing. The student can verify the soundness of this inference on a monochord; or, with care, it is possible to do so on a pianoforte string. To experiment with a pianoforte string a small weight should be placed on the corresponding digital. A string wrapped with wire will not give a satisfactory result. As Donkin remarked,[1] the wire covering appears to have

[1] *Acoustics*, W. F. Donkin, p. 95.

the effect of partially restoring the even partials which would otherwise be eliminated by plucking the string at its middle point. We must therefore choose a note which sounds on more

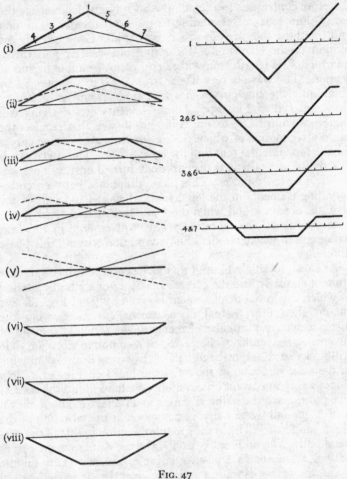

FIG. 47

than one string. To avoid resonance effects, the other strings which sound the same note should be silenced. Pluck the string at its middle point. This will cause it to vibrate, for the reason given above, without the octave harmonic, and in fact without any even partials. No node will be formed at the middle point. Now touch the string at its middle point immediately after plucking it there: this will destroy all vibrations except any which have a node there. But as there will be none with a node there the string will be silenced. If a string be struck or plucked one-third of its length from one end, the 3rd, 6th, 9th, 12th, &c., partials will be lacking. Similarly for other points. This will be important in considering later the points at which a pianoforte string is struck.

The actual displacement curves for particular elements of the string, numbered 1 to 7, are shown on the right-hand side of Fig. 47. It will be observed that they are not sine curves. Displacements of the string in other shapes could be worked out and the displacement curves could be drawn for particular elements of the string. An interesting one to work out is a string plucked one-ninth of its length from one end. This produces a sound containing all the partials except the ninth, eighteenth, twenty-seventh, and so on. The vibration is consequently interesting to compare with that in Fig. 44 which produces a sound in which all the odd partials but only the alternate even partials, beginning with octave harmonic, are present. If he wishes, the student may work out the vibrations of a string released from the position shown in (ii) in Fig. 38, which represents a string plucked simultaneously at points a quarter of the length from each end. It will be found by repeated trials that the displacement curves of a vibrating string are never all sine curves unless the initial displacement of the string itself is a sine curve.

A pianoforte string is usually struck by the hammer from a seventh to a ninth of the way along it. This practically stifles all the harmonics above the sixth partial. The top strings are struck still nearer their end, for reasons explained below. The general character of the vibrations produced by a blow from the hammer can be inferred from that of a plucked string; but they differ for two reasons from those of a plucked string. First, the hammer has not a sharp edge such as would give a sharp angular displacement where it struck the string. The angular displacement would be rounded because the hammer makes contact for some little distance. Second, the hammer is not quite hard, it has a fair degree of elasticity. The actual striking therefore occupies an appreciable interval, during which the displacement can begin to run along the string. This will further round off any angularities in the vibration curve we should obtain by plucking the string. The higher harmonic overtones are therefore much softened as compared with those of the plucked string. This helps to make the tone of the piano much rounder than that of the harpsichord. The top strings are so much shorter than the middle ones that they are comparatively stiff. This makes their natural harmonics weaker than those of longer strings.[1] The student who has worked out the vibration curves of a string plucked one-ninth of its length from one end, will realize that the higher harmonics are encouraged by striking it nearer one end with a pianoforte hammer. The selection of a striking point nearer the end of the top strings also causes the narrow hammer to rebound very quickly; consequently the contact is too brief and extends over too short a length of string to smooth out the angularities of the vibration curve as much as they would be smoothed out by a softer and broader hammer striking farther from the end. These factors help to compensate for the relatively greater stiffness of the short strings, and impart the necessary brilliance to their tone.

The behaviour of the violin string when played pizzicato is that of a plucked string. Bowing excites quite a different motion; and the action of the bow seems to be as follows. The bow, made tacky with rosin, seizes an element of the string and draws it to one side with uniform speed. Eventually it loses its hold, and the string swings rapidly back with a greater but practically uniform speed (there is a slight fluttering in it) to a point on the other side of its equilibrium position. It is there seized again by the bow and the same performance is repeated.

The actual vibration of the violin string was recorded by Helmholtz by means of very elegant and delicate experiments. He showed that the string, at any moment, has the form of two straight lines meeting at an angle. He confirmed his conclusions by mathematical analysis, which showed that the apex at which the two strings meet, travels round two practically circular arcs as shown in Fig. 48 by the arrows.

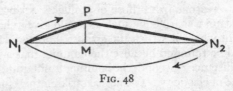

FIG. 48

If P is the apex and PM is perpendicular to $N_1 N_2$, M will move backwards and forwards with uniform speed between N_1 and N_2. The vibration will obviously result in the ear hearing an extended succession of partials.

[1] See p. 126.

We began this chapter by assuming some of the results which mathematicians obtain through calculating the motion of vibrating strings. We have thereby been able to find out *how* strings vibrate. We have still to indicate the reasons, disclosed by mathematical analysis, *why* they vibrate as they do. It was stated in the Introduction that the science of dynamics, which deals with forces, masses, velocities, and accelerations, is required to determine the laws of vibrating bodies. An indication of how it does so in the case of a perfectly uniform and flexible string is given in Appendix X. It is there shown, by dynamics, that a deformation travels along a perfectly uniform and flexible string without loss of shape and with uniform velocity. This was the assumption we made in considering the reflection of a wave at the fixed end of our garden hose. The observations which the student may have made of the effect of sending a wave along a garden hose would be too crude and inexact as an experimental basis for the laws of vibration of strings. But they will have served to prepare him to accept the conclusion, reached by dynamics, in Appendix X. The conclusion that a deformation travels along a perfectly uniform and flexible string without loss of shape and with uniform velocity is the key to the whole problem. From it follows the deduction that a vibrating string behaves as if disturbed by a stationary wave produced by reflections at each end, the ends becoming nodes. Mathematical calculations of the behaviour of pianoforte and violin strings are given at the end of Helmholtz's work. They express vibrations in terms of Fourier's series which, as explained in Appendix V, have their counterpart in the analysis which the ear makes of a sound wave. We may conclude from all this that the deductions we have reached graphically from our assumptions have a sound basis in dynamical science.

A summary of the laws of vibration of strings may well be recorded. We have already observed that a heavy string vibrates more slowly than a light one, and one tightly stretched more rapidly than one under less tension. This is the result of the differences made by the weight and the tension of the string to the speed with which a wave runs along it. In Appendix X the speed is calculated at which a wave would run along a string whose weight and tension are known. From the result it follows that the frequencies of vibration of strings of the same lengths vary inversely as the square root of the weight per unit length, and directly as the square root of the tension. We know already that the frequencies of vibrating strings vary inversely as their lengths. These are the laws of vibration of strings, first investigated scientifically by Mersenne, as explained in a footnote on p. 7, and later expressed by Brook Taylor[1] in a mathematical formula, which was equivalent to the formula given at the bottom of p. 167.

Thus we have at last reached the point which scientific textbooks on sound sometimes take as a starting-point, namely, the rate of vibration of a string as established experimentally with a monochord. This completes our task. If the student has worked through the last five chapters with squared paper, drawing his own diagrams whenever possible, he will have acquired an exact knowledge of certain acoustical effects of importance to music which will equip him to read with a critical appreciation a standard work on the subject. Even if he pursues his studies no further, he will have formed a clear conception of the factors which determine the shape of the wave representing, to his ear, any music to which he may be listening. The whole sound of an orchestra reaches his ear at any moment as a single wave. It excites vibrations which the ear analyses, instantaneously and continuously, into an ever varying series of harmonic motions which can be represented by sine curves; and these it arranges in such a way as to convey to him the sense of the quality of the various instruments and the notes they are playing. This explains why it is possible to record orchestral music by the impressions on the bottom of the groove in a gramophone record which actuate the vibrations of the diaphragm through the needle.

To attain an understanding of this and similar matters of interest to musicians the student has been urged to draw diagrams on squared paper for himself, so as to explore and master the various matters we have discussed. Representation of vibrations graphically will alone give to the non-mathematical student a clear grasp of their actual nature. If the process of drawing them calls for both time and patience, he may reflect that it would take him far longer to acquire that mastery of mathematical technique which would provide him with other means of attaining the same end. He may reflect, too, that by using graphical methods instead of solving differential equations and employing Fourier's series he is following in very illustrious footsteps. In writing his *Principia*, Newton was unable to use mathematical technique of the kind described, for it had still

---

[1] Brook Taylor's name is made familiar to all mathematical students by 'Taylor's theorem'.

to be invented. Yet, by purely geometrical deduction, he was able to demonstrate that the laws of planetary motion, laid down from direct observation by Kepler, followed from the law of gravitation he had announced; and thus he established an unassailable presumption of its truth. The time which the student may have spent in drawing sine curves will have enabled him to exhibit properties of sound waves in a manner which shows agreement with observed facts; and he will thus have established for himself an unassailable presumption of the truth of the statement that a pure tone is represented by a longitudinal vibration in air which takes the form of harmonic motion.

APPENDIX I

TABLE I

Logarithms of numbers from 1 to 20 to five places of decimals

log	1	0·00000	log 11	1·04139
log	2	0·30103	log 12	1·07918
log	3	0·47712	log 13	1·11394
log	4	0·60206	log 14	1·14613
log	5	0·69897	log 15	1·17609
log	6	0·77815	log 16	1·20412
log	7	0·84510	log 17	1·23045
log	8	0·90309	log 18	1·25527
log	9	0·95424	log 19	1·27875
log	10	1·00000	log 20	1·30103

TABLE II

Sines of angles from 0° to 90° to five places of decimals

Increasing by a quarter of a right angle		Increasing by tenths of a right angle	
sine 0°	0·00000	sine 0°	0·00000
sine 22½°	0·38268	sine 9°	0·15643
sine 45°	0·70711	sine 18°	0·30902
sine 67½°	0·92388	sine 27°	0·45399
sine 90°	1·00000	sine 36°	0·58778
		sine 45°	0·70711
		sine 54°	0·80902
		sine 63°	0·89101
		sine 72°	0·95106
		sine 81°	0·98769
		sine 90°	1·00000

TABLE III

Antilogarithms required to calculate, to five significant figures, the frequency of the notes of an octave in equal temperament beginning with c′ as 256

$$2·45841 = \log 287·35$$
$$2·50858 = \log 322.54$$
$$2·53367 = \log 341·72$$
$$2·58384 = \log 383·57$$
$$2·63401 = \log 430·54$$
$$2·68418 = \log 483·26$$

TABLE IV

Frequencies, to five significant figures, of the notes of an octave beginning with c' as 256

Pitch	Just Intonation	Equal Temperament
c'	256	256
d'	288	287·35
e'	320	322·54
f'	341·33	341·72
g'	384	383·57
a'	426·67	430·54
b'	480	483·26
c"	512	512

If c' were given a higher frequency than 256 the frequencies of the other notes could be calculated by simple proportion from those given above. Thus, if c' were given a frequency of 258·7 the other frequencies would all be increased in the ratio 258·7 : 256.

The figures in the second decimal place in Table IV are somewhat of an arithmetical refinement; but so accurate is the method of tuning by beats, explained in Chapter V, that a skilled tuner could produce an octave of the scale more accurately tempered than that which would be exhibited by a column of frequencies calculated to only four significant figures.

While it is possible by careful tuning to fix intervals of an octave and of a just or tempered fifth with surprising accuracy, it should be remembered that in the performance of music which is sung or played with free intonation, such a degree of accuracy is seldom attained and is just as seldom of musical importance. For reasons explained in Chapter III and Chapter V, this is true more particularly of what the musician describes as either unessential notes or notes of essential discords. Writers on scales are apt to mislead when they calculate, with extreme accuracy, what they take to be a theoretical size for the intervals which occur in various chords catalogued in books on harmony. This they sometimes do in terms of the hundredth part of a semitone of equal temperament, which Ellis called a cent. As a mistuned unison in the middle of the treble clef an interval of one-hundredth part of a semitone would produce beats at the rate of about one every four seconds. A name given to such an interval tends to suggest and foster the idea that it carries some musical meaning, whereas the original intention of the cent was to provide a substitute for the use of logarithms for measuring intervals. Helmholtz himself used no such term. The errors which carefully tuned temperaments introduce into the scale of keyed instruments of fixed intonation are in quite a different case from theoretical inexactnesses of many of the notes used, quite artistically, in the performance of music on instruments whose intonation is not so rigidly fixed. The niceties of 'intonations' have an arithmetical fascination which must be resisted. To yield is to miss the full significance of Helmholtz's work on consonance, dissonance, definition and the effects of mistuning, and to fall a victim of the 'miasma' against which Stanford warned us, a miasma caused by supposing that the technique of music is mainly comprised in the classification of the harmonic relations of notes viewed vertically on the score. (See *Musical Composition*, Stanford, Chapter II.) Helmholtz gives much the same warning in another connexion, and from a different point of view: 'We must distinguish carefully between composers and theoreticians. Neither the Greeks, nor the great musical composers of the sixteenth and seventeenth centuries, were people to be blinded by a theory which their ears could upset.'

A reader who has mastered Chapter V will see that Stanford's advice to learn harmony through counterpoint is good science as well as sound musicianship. Counterpoint, with its horizontal outlook, introduces into technical study the effects of both the time factor in music and the physiological limits of the ear.

APPENDIX II

Table

Showing the intervals between the first note and each successive note of an octave of just and of tempered intonations, as measured by their logarithms

Degree of Scale	Just Intonation Logarithms of Intervals	Mean-Tone Tuning Logarithms of Intervals	Equal Temperament Logarithms of Intervals
1st	..	..	..
2nd	0·05115	0·04845(5)	0·05017
3rd	0·09691	0·09691	0·10034
4th	0·12494	0·12629	0·12543
5th	0·17609	0·17474	0·17560
6th	0·22185	0·22320	0·22577
7th	0·27300	0·27165	0·27594
8th	0·30103	0·30103	0·30103

To obtain a picture of the relatively small differences between the corresponding intervals of each octave, it is interesting to make a diagram on squared paper showing the logarithms of these intervals side by side in the manner in which the right-hand column of the diagram in Chapter I, p. 19, was constructed.

APPENDIX III

(See page 69)

It was shown in Chapter VI that sound waves are propagated through the air by a series of alternate compressions and expansions.

Robert Boyle (1627–91) enunciated a rule, commonly called Boyle's law, for ascertaining the change caused in a given volume of air, or other gas, by a change of pressure. This states that the volume varies inversely as the pressure if the temperature is kept constant; and if the volume is v and the pressure p, this rule can be expressed by the equation

$$p \times v = \text{constant}$$

Newton (1642–1727), assuming this rule to apply to the compressions and expansions caused in air by sound waves, used it to determine theoretically the velocity of sound in air. His result disagreed with the value obtained experimentally, and he was led to propose an explanation in terms of the particles of air which we now know to be unnecessary, and inconsistent with the results of all other physical research bearing on the issue.

Laplace (1749–1829) traced the discrepancy to something not known to Newton, and his correction is important. To-day we know that when a bicycle is pumped up the valve and the nozzle of the pump become hot. It is not so generally known that when the valve plug is pulled out and the air is allowed to escape the valve becomes cold, but it is equally true. Compression of air heats it; expansion cools it. The compressions and expansions produced in air by a sound wave cause alternate heating and cooling effects which occur too rapidly to be dispersed through the air. The compressions and expansions are then said to be *adiabatic* to distinguish them from *isothermal* compressions and expansions in which the temperature is constant and to which Boyle's law applies. Now when a volume of air is heated and prevented from expanding it requires less heat to raise its temperature by one degree than is the case if it is allowed to expand, for example along a tube leading from the vessel containing it. This is natural, for if the air were allowed to expand along the tube, it would do work in overcoming the resistance of the external air, for example by pushing a column of mercury along the tube. To do so it would absorb more heat. The energy required to heat the air when it is allowed to expand, exceeds the energy required to heat it when not allowed to expand, by the energy represented by this work. For gases like oxygen and nitrogen the ratio between the quantities of heat required in the two cases is constant at different temperatures. It is denoted by the sign γ, which is called the *ratio of the specific heats of the gas*; and for gases like oxygen and nitrogen γ is found to be 1·41.

Laplace showed that for adiabatic compression and expansion Boyle's law needed to be restated as shown by the equation

$$p \times v^{\gamma} = \text{constant}$$

This correction altered the position completely. Suppose that p is increased by a *small* amount δp, and that the consequent *small* decrease in v is represented by δv. It follows that

$$(p+\delta p)(v-\delta v)^{\gamma} = \text{constant} = pv^{\gamma}$$

Expanding by the binomial theorem and regarding the square and higher powers of δv and the product of δp and δv as negligible, this gives

$$(p+\delta p)\left(1-\gamma\frac{\delta v}{v}\right)v^{\gamma} = pv^{\gamma}$$

or

$$p+\delta p-\gamma p\,\frac{\delta v}{v} = p$$

or

$$\frac{\delta p}{\delta v} = \gamma\frac{p}{v}$$

Hence for given values of p and v we see that $\dfrac{\delta_1 p}{\delta_1 v} = \dfrac{\delta_2 p}{\delta_2 v}$, where δ_1 and δ_2 are used for different small displacements.

It is very important to note that this relationship is not true if δp and δv are not so small that $\delta^2 v$ and $\delta p\delta v$ may be neglected. It explains why the conclusions which follow do not apply if the sound is loud, and in particular why these conclusions do not preclude the possibility that a combination tone having an actual, objective, existence may be produced by two generators sounding loudly (see Chapter IV). The mathematical proof that such a combination tone not only may, but will, be produced is beyond the scope of this book.

The equation
$$\frac{\delta_1 p}{\delta_1 v} = \frac{\delta_2 p}{\delta_2 v}$$

shows that for small displacements, air behaves like an elastic solid to which Hooke's law applies, namely that, within the limits of elasticity, the amount of the distortion is proportional to the distorting force.

Now turn to Fig. 19 in which P is supposed to move round a circle with uniform speed. Let the angular velocity of OP, which is constant, be ω. The string's pull exerts a force on the weight P which causes a constant change of its direction of motion; such a change is an acceleration and its direction is always along PO. It is shown in books on dynamics that this acceleration is $\omega^2 r$, where r is the length of PO. Consequently the component of this acceleration parallel to CD, which is the acceleration imposed on N, is $\omega^2 r \cos P\hat{O}N$ or $\omega^2 NO$ and takes effect in the direction of O. This, by Newton's laws of motion, means that N may be regarded as acted on by a force equal to NO and directed to O; in other words exactly the ⌄rce which would be exerted on N if it had to obey Hooke's law. Since it does so *for small displacements* it follows that if the sound is not too loud the motion it will excite in the air must be *harmonic motion* or a *pendular vibration*.

Observe that the acceleration is in terms only of ON and ω^2: the rate at which N will oscillate will be independent of the size of the circle so long as ω is constant. The length of CO is the amplitude of the vibration, the size of ω determines its frequency. Pitch depends on frequency. For a given pitch, loudness depends on amplitude, that is, the maximum displacement to the right or left of the equilibrium position. Were the oscillations of N not independent of their amplitude it would be necessary for notes to alter in pitch if they were altered in loudness. (For the effect of combined alteration of frequency and amplitude on loudness, see Appendix V.)

APPENDIX IV

(See page 74)

In Chapter VI, at the end of the discussion of a progressive wave, reference was made to the associated velocity and condensation curves, which are both represented in (vi) of Fig. 20.

We may observe that the linear velocity of the motion of P in Fig. 19 is ωr, and its component in a direction parallel to OC is $\omega r \cos P\hat{O}A$ which is the same as $\omega r \sin (90° + P\hat{O}A)$. The changes of velocity may be represented graphically by a sine curve, but it will differ in phase from the associated displacement curve. It will be necessary to adopt a convention: velocity in the direction in which the sound is travelling will be shown in an upward direction, velocity in the direction opposite to that in which the sound is travelling will be shown in a downward direction.

Further, consider the changes in condensation and rarefaction—which are the direct effect of the displacement. These changes, like those of displacement, may therefore be represented by a sine curve, but it will be different in phase from that of the associated displacement curve. Again a convention is necessary; condensations will be shown in an upward direction, rarefactions in a downward direction.

Reference to Fig. 20 will show that, when the condensation at a given point is a maximum, a particle of air at its centre is in its equilibrium position, that is, has zero displacement but is moving forward with maximum velocity. When the condensation has been passed on and there is neither condensation nor rarefaction at this point, the displacement to the left is a maximum but the velocity is zero. After an equal interval of time the rarefaction is a maximum; the particle then has no displacement but is moving with maximum velocity in a direction opposite to that in which the sound is travelling. And so on. It will be clear therefore that the same sine curve can be used to represent the condensation and the velocity, but it will differ by a quarter of a phase from the associated sine curve which represents the displacement. Consequently, when the sine curve shown in (v) represents the displacement, the sine curve shown in (vi) will represent both the condensation and the velocity. It must not be forgotten that these *associated* curves are only diagrammatic and are used only to make the changes in the *actual* displacement, condensation, and velocity more readily appreciated by the eye.

APPENDIX V

(See pages 76 to 79)

IN Chapter VI it was explained how the complex wave form which represents the vibration of a particle of air when affected by simultaneous vibration due to two or more pure tones could be built up, graphically, by compounding the sine curves representing these simple vibrations. The converse process of analysing a complex wave form into its component harmonic vibrations can be performed by mathematicians if the complex wave form is periodic. A theorem of very wide application, due to Fourier, provides the necessary tool. This theorem states in effect that, with certain exceptions not possible for sound waves in air, any periodic vibration of wave-length λ can be expressed mathematically by harmonic vibrations represented by sine curves of wave-lengths $\lambda, \dfrac{\lambda}{2}, \dfrac{\lambda}{3}, \dfrac{\lambda}{4}$, and so forth, with suitable amplitudes. The amplitude of any component harmonic vibration is represented by the coefficient of the corresponding term of the Fourier's series as written down by the mathematician.

In Chapter VI the complex vibrations produced by combining pure tones at intervals of a perfect fifth, a perfect major third, and a minor tone were built up graphically, and they were found to be periodic. It was stated that analysis of the result could not get out of it anything more than was put into it. If we denote by λ the wavelength of the compounded periodic curve in each case, Fourier's theorem would apparently produce as an analysis of this curve, a series of sine curves of wave-length $\lambda, \dfrac{\lambda}{2}, \dfrac{\lambda}{3}$, &c. Why then do we state that analysis of the vibration representing a fifth between pure tones gives only sine curves of wave-length $\dfrac{\lambda}{2}$ and $\dfrac{\lambda}{3}$, of that representing a major third only those with wave-lengths $\dfrac{\lambda}{4}$ and $\dfrac{\lambda}{5}$, and of that representing a minor tone only those with wave-lengths $\dfrac{\lambda}{9}$ and $\dfrac{\lambda}{10}$? The answer is that the coefficient of all other terms of the Fourier's series is nought, or that the corresponding sine curve has a zero amplitude, so that the corresponding tone does not exist. Thus in the case of the fifth the terms containing the series of wave-lengths $\lambda, \dfrac{\lambda}{4}, \dfrac{\lambda}{5}$, &c., all have nought for their coefficient.

(See pages 83 to 85)

The footnote on p. 74 called attention to the fact that by selecting the same amplitudes for two tones a fifth apart we should be representing the upper one as louder than the lower one. The distinction between loudness and intensity of sound is given in Chapter VII, p. 93. Regarding the rate at which energy is transmitted across unit area of a plane parallel to the front of a progressive wave as the mechanical measure of the intensity of sound, Rayleigh found for the intensity of a pure tone the value:

$$2\pi^2 \rho a \frac{\xi^2}{\tau^2}$$

where ρ is the density of the medium, a the velocity of sound in it, ξ the amplitude and τ ($= \lambda/a$ or $1/n$) the periodic time[1]. It follows that the intensity varies directly as the square of the amplitude, and inversely as the square of periodic time (or of the wave-length). In other words the intensity varies directly as the square of the product of the amplitude and the frequency (n).

Consequently if we wished to represent the two tones in Fig. 22 as having the same intensity we should have to give the upper one two-thirds the amplitude of the lower one. The product of the amplitude and the frequency would then be the same in each case. Similarly in Fig. 23 the upper tone would need an amplitude which was four-fifths that of the lower one if they were to be represented as having the same intensity. The student who cares to make the necessary calculation, quite simple with a slide-rule, may be interested (say on a second reading) to re-draw one of these figures for tones of equal intensity. The character of the resultant curve would be unchanged; its proportions would be slightly different from those represented in our figure as drawn.

It is interesting to apply Rayleigh's formula to the amplitude of the curves we selected for Figs. 25a, 25b, and 25c. The intensities of the fundamental, the octave harmonic, and the twelfth in Fig. 25b are in the ratios 36 : 64 : 81. The outstanding feature of the trace, as recorded by Miller's phonodeik, of a flute played forte, is that the two kinks in the descending limb of the curve actually bend upwards. This upward bend is perhaps exaggerated in Fig. 25b; but trial graphs will show that, to produce it, the third or some higher partial tone must have had considerable intensity. Miller's actual analysis showed a third partial with about the same intensity as the fundamental and a fourth partial half as loud again. (See *Science of Musical Sounds*, Miller, p. 193.)

In (iv) of Fig. 25c the intensities of the fundamental and the octave harmonic are in the ratio 25 : 4. It was explained on page 85 that to reproduce closely the trace, as recorded by Miller's phonodeik, of a flute played piano the amplitude of the fundamental in (iv) of Fig. 25c should be increased and that of the octave harmonic reduced, which would increase considerably the ratio of the intensity of the fundamental to that of the octave harmonic. The flute would then be represented as sounding its fundamental as a nearly pure tone.

[1] *Theory of Sound*, 2nd ed., § 245.

APPENDIX VI

(*See pages* 98 *to* 102)

THE frequency of a resonator has been calculated by Rayleigh, who gives for it the formula $\frac{a}{2\pi}\sqrt{\frac{c}{S}}$, where a is the velocity of sound in air, c the conductivity of the mouth, and S the volume of the resonator.[1] The conductivity of the mouth depends mainly upon its size.

The experiments described in Chapter VIII show us how the frequency depends on the size of the mouth of the bottles used, and on the volume of the bottle. It is not difficult to show that the frequency varies inversely as the square root of the volume. To give the following experiment a quantitative value it is useful, for measuring the water, to employ a measuring vessel, such as is sometimes used in the kitchen to measure milk; or a medicine bottle marked in tablespoons will serve equally well. If a bottle which gives a good note is half filled with water the pitch of its note will be found to have risen by a tritone or an imperfect fifth—intervals which are the same in equal temperament, when they are each exactly half an octave. We know therefore that the frequency of the note of the bottle has been increased by a figure equal to $\sqrt{2}$, for the ratios of two intervals, each equal to that by which the note has increased, when multiplied must be 2, which is the ratio of an octave. If we fill the bottle three-quarters full of water, so as to reduce the volume of the air to one-quarter of its original size, the note will rise an octave. In mathematical language the frequency varies inversely as the square root of the volume.

In Chapter VIII mention was made of a simple experiment to show how the inharmonic first overtone of a tuning-fork may be exhibited. For this experiment it is necessary to have a vice and a piece of steel wire about a yard long of, say, 10 to 12 wire gauge. Fix one end of the wire in the vice, and make the wire vibrate as a whole as shown in (i) of Fig. 49.

Now hold the wire between the finger and thumb of one hand about a fifth, or a little more, of the way along it, measuring from the free end. With a finger of the other hand bend the wire into the form shown in (ii) and release both hands simultaneously. The wire will vibrate as shown in (ii) of Fig. 49, and a little experiment will find the exact spot N which will eliminate as nearly as possible all the vibrations of the form shown in (i). The increase in the rate of vibration compared with that shown in (i) is very marked; it will be about $6\frac{1}{4}$ times as rapid. If the distance of N from the free end of the wire is carefully measured, it will be found to be a little more than a fifth

---

[1] *Theory of Sound*, 2nd ed., § 304.

of the whole length of the wire measured to the vice. These figures explain why the first overtone of a tuning-fork, each prong of which may be compared to our wire, is not a harmonic overtone of the fundamental save in fortuitous and very exceptional circumstances. It is possible to make the wire vibrate with two nodes, like *N*, when

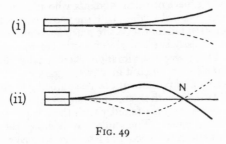

FIG. 49

its motion will represent the second inharmonic overtone, but this is not easy to manage.

Finally, it may be useful to call attention to the difference in the behaviour of a resonator of bottle shape and that of a cylindrical jar the opening of which is not constricted. Neglecting the correction for open end, explained in Chapter IX, the natural mode of vibration of a cylindrical vessel depends only on its length and not on its volume; for its behaviour is similar to that of a stopped organ pipe. The air all down the vessel is set in vibration. With the bottle the energy of the movement of the air is confined mainly to the neck and its immediate neighbourhood.

APPENDIX VII

(*See pages* 70 *to* 73, *and* 106 *to* 111)

IT is a common experience that students who have not been made familiar, in mathematical studies, with the use of curves for graphical representation often fail thoroughly to grasp and therefore to remember just what an associated displacement curve means, and that it represents, not the transverse vibration which it exhibits, but a longitudinal vibration in air which it greatly exaggerates. In truth, associated displacement curves are apt to be confusing to any one till he makes himself familiar with them. But they present no real difficulty: all that is needed to become familiar with them is patience, a pencil, and some squared paper. The student who draws them and combines them for himself will sooner or later acquire a complete mastery of them even though his knowledge of mathematics is negligible. He will then find that he possesses a weapon with which he can attack many of the acoustical problems which interest musicians. For that reason we began in Chapter VI by drawing, on a very exaggerated scale, the actual longitudinal displacements of particles of air when disturbed by a sound wave, and from them we derived the associated displacement curve. For the same reason, in Chapter IX we ended the study of stationary waves in air by drawing, on a very exaggerated scale, the actual longitudinal displacements of particles of air in Fig. 29. In the diagrams of the organ pipe we used arrows to exhibit the direction of the longitudinal displacements on both sides of a node due to the condensation or rarefaction there. In books on acoustics intended for science students the associated displacement curves are often drawn inside the organ pipe. The picture lingers in the memory of the non-mathematician when its explanation is forgotten. The result is particularly unfortunate when the picture is confused with that of a vibrating string sounding its fundamental tone. Hence one advantage of reserving till the last chapter of this book the study of vibrating strings. The student is strongly urged to carry in his memory the horizontal arrows in Figs. 31 to 34 as his last impression of organ pipes. They will help him to remember that at a node in an organ pipe the condensation and rarefaction is always a maximum, and to think of the motion of the particles of air as being longitudinal.

Those who are quite familiar with graphical representation in mathematics may confirm their understanding of the nature of stationary waves by drawing a complete series of associated displacement curves, velocity curves, and condensation curves as shown in Fig. 50. The first group will be when all the particles are in their equilibrium position; the second when they are all displaced to the maximum amount; the third when they are all in an intermediate position. The displacement curves we have already drawn: they are

ASSOCIATED DISPLACEMENT, VELOCITY, AND CONDENSATION CURVES
FOR STATIONARY WAVES

N represents a Node, A an Antinode

Position I
(Corresponding to (i) of Fig. 28 and 1 of Fig. 29)

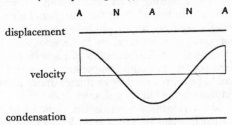

Position II
(Corresponding to (v) of Fig. 28 and 5 of Fig. 29)

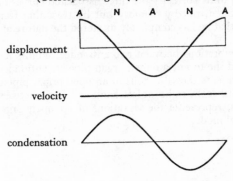

Position III
(Corresponding to (ii) of Fig. 28 and 2 of Fig. 29)

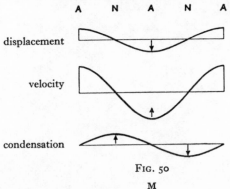

FIG. 50

M

to be found in (i), (v), and, say, (ii) of Fig. 28 respectively. We will nevertheless repeat them in Fig. 50, with the vertical scale reduced and the horizontal scale increased, as is convenient for comparison with the associated velocity and condensation curves. There are three other diagrams which can usefully be drawn for intermediate positions, namely those corresponding to (iv) and (vi) and to what would be (viii) of Fig. 28, and in two of them the particles of air are moving towards instead of away from their equilibrium positions. It is necessary to be clear about the conventions to be adopted. In Fig. 28 we used the convention that displacements of particles to the right of their equilibrium positions in both component waves and in the stationary wave were to be represented by ordinates drawn above the centre line, displacements to the left by ordinates drawn below it. Similarly we must now show velocities towards the right by ordinates drawn above the centre line and velocities to the left by ordinates drawn below it. Condensations will be shown by ordinates drawn above the centre line, rarefactions by ordinates drawn below it. When the student can draw correctly the three associated curves for all the positions indicated, not by memory but by reasoning from Fig. 29, he will know that he has completely mastered the nature of a stationary wave.

Fig. 50 represents vibrations between two antinodes. Points just outside the mouth and the top of an open organ pipe are antinodes: Fig. 50 therefore represents the vibrations of an open organ pipe in its octave mode; and either the right- or left-hand half, lying between consecutive antinodes, represents the vibrations of an open organ pipe in its fundamental mode.

APPENDIX VIII

(See pages 115 and 122)

Correction for Open End.

RAYLEIGH observed[1] that it was natural that a much larger correction (for deficient openness) should be required at the lower end of an organ pipe, which was partially closed at the mouth, than at the upper open end. Cavaillé-Coll showed from actual experiments with organ pipes that the total correction for both ends in an open organ pipe was of the order of 1·6 times the diameter. That an additional correction is necessary at a partially open end is indicated by, and suggests an explanation of, a method of tuning open flue pipes. To lower the pitch, a hollow cone is pressed over the open end which closes the opening a little. To raise the pitch, a solid cone is pressed into the open end which enlarges the opening a little. Many metal organ pipes have two strips of metal standing out on either side of the mouth and at right angles to the lip. If bent inwards they make the opening smaller, and increase the correction required for deficient openness: that is, they lower the pitch. If bent outwards they reduce the correction and raise the pitch. If the organ pipe is considered as a coupled system consisting of a resonator sounded by blowing across the mouth, the effect of bending the metal strips at the mouth inwards is to reduce the conductivity of the mouth of the resonator, that is, to lower its pitch.

In the flute the effective length of the pipe obtained by uncovering a finger hole is increased if the hole is made smaller. This effect has an analogy with the correction for 'deficient openness', and use is made of it in placing the holes in positions convenient for fingering.

Effect of Temperature.

It can be shown mathematically that the velocity of sound in a gas is given by the equation

$$V = \sqrt{\gamma \frac{p}{\rho}}$$

where V is the velocity, γ the ratio of the specific heats (see Appendix III), p the pressure, and ρ the density of the gas when undisturbed by a sound wave. This value of V contains no factor for either the amplitude of the vibration or its frequency, and for small vibrations to which alone it applies, for reasons explained in Appendix III, V is independent of the pitch or the loudness. If now v be the volume of unit mass of the gas, $\frac{p}{\rho}$ is the same as pv; and we know by Boyle's law that this is constant for a given gas if the temperature is not changed.

[1] *Theory of Sound*, 2nd ed., § 322*a*.

It follows that changes of barometric pressure have no effect on the velocity of sound or the pitch of an organ. But, by what is known as Charles's law, when the temperature changes pv equals κT, where κ is a constant and T is the absolute temperature, or the centigrade temperature plus 273°. Consequently $V = \sqrt{\gamma \kappa T}$ and the velocity is proportional to the square root of the absolute temperature. The effect of an increase of temperature is to make the wave motion up the organ pipe, which with its reflection at the top creates the stationary wave, more rapid. This is by no means compensated for by the expansion of the pipe with increase of temperature. The effect is to make the pipe behave as a shorter pipe would do at the original temperature. A rise of temperature therefore causes the pitch of flue pipes to rise.

With a rise of temperature the metal reed of a reed pipe in an organ becomes longer and slightly less stiff. This reduces the frequency of its natural vibration. It is then able to 'force' a reduction of the natural frequency of vibration of the air in the pipe. Reed pipes therefore become sadly out of tune with the flue pipes when the temperature changes materially.

The effect of a rise of temperature on the instruments of the orchestra is similar to, though less than, the effect on a flue pipe in an organ; in the average it is about half as much. Changes of temperature in a concert-hall have less effect on small instruments, easily warmed by the player's breath, than on large ones, the temperature of which cannot be kept so steady by the player's breath.

APPENDIX IX

(See page 139)

IN Chapter X we showed how we could always discover the vibrations of a string vibrating freely from any given displacement. The wave form is drawn as two progressive waves proceeding in opposite directions, and the result is a stationary wave representing the vibration of the string. This method of finding the vibration can be used, as we have seen, when we do not know, as the mathematician knows, the sine curves which built up the original displacement. But if we know these sine curves, as we did in the case of the string displaced as shown in Fig. 41, we can employ an alternative method. We could work out the separate harmonic vibrations from their sine curves and then combine them. We should then simply be reversing the process which the ear adopts. We could for example apply this method to calculate the vibrations produced by the displacement shown in Fig. 41. If the student does so he will find that he obtains exactly the same result as is shown in Fig. 46. This will convince him, if he has any lingering doubts, that our procedure is sound.

If he wishes to do this he will find the following table useful in calculating the positions at the beginning and end of four periods each equal to one sixteenth of a complete vibration. The figures at the head of the table are the sines used in plotting sine curves as shown in Fig. 21 of Chapter VI.

	Posi-tion	1·0	0·95	0·89*	0·8	0·7	0·45	0·3
Funda-mental	1	1·0	0·95	0·89	0·8	0·7	0·45	0·3
	2	0·92	0·87	0·82	0·73	0·64	0·41	0·28
	3	0·71	0·675	0·63	0·57	0·50	0·32	0·21
	4	0·38	0·36	0·34	0·31	0·27	0·17	0·11
	5	0·0	0·0	0·0	0·0	0·0	0·0	0·0
Octave	1	0·5	0·475	0·45	0·4	0·35	0·22	0·15
	2	0·35	0·34	0·31	0·28	0·25	0·16	0·10
	3	0·0	0·0	0·0	0·0	0·0	0·0	0·0
	4	−0·35	−0·34	−0·31	−0·28	−0·25	−0·16	−0·10
	5	−0·5	−0·475	−0·45	−0·4	−0·35	−0·22	−0·15

* 0·89 is the sine of 63°, and is given to define the curves at this point. Other sines can be added if desired from Table II, Appendix I.

APPENDIX X

(See page 145)

THE following elegant and elementary mathematical determination of the way in which and the velocity with which a transverse deformation travels along a perfectly uniform and flexible string is due to Tait. Imagine the string to be at rest and to be enclosed in a smooth tube which is deformed transversely in the shape of any given deformation. A short element of the string PQ, Fig. 51, having a radius of curvature r, is maintained in equilibrium by the tension T at each end of the element and the resultant reaction of the tube between P and Q. If now the string be made to move along the tube with a velocity v, the element PQ is also subject to a centrifugal force which is shown in books on dynamics to be $m\omega^2 r$, where ω is the angular velocity of a point in the element PQ and m is the mass of PQ. This equals $m\left(\dfrac{v}{r}\right)^2 r = \dfrac{mv^2}{r}$.

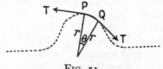

FIG. 51

Now suppose the velocity with which the string is running through the tube to be such that there is no pressure on the tube from the element PQ: in other words that, by its centrifugal force, it is just kept from pressing on the tube. The resultant reaction of the tube on the element PQ of the string is now zero; and the element PQ is now acted on by the tensions at the end and the centrifugal force alone. If θ be the small angle between the tangents at P and Q, it follows that $2T \sin \tfrac{1}{2}\theta = \dfrac{mv^2}{r}$, for $2T \sin \tfrac{1}{2}\theta$ is the resultant of tensions each equal to T at each end of the element.

But since θ is small $\sin \tfrac{1}{2}\theta = \tfrac{1}{2}\theta$, and consequently $T\theta = \dfrac{mv^2}{r}$.

Now suppose that the mass of unit length of the string is ρ. It will follow that $m = \rho r \theta$.

It follows that
$$T\theta = \rho\theta v^2$$
or
$$T = v^2\rho$$
or
$$v = \sqrt{\dfrac{T}{\rho}}$$

This contains no element representing r, and it therefore follows that when the velocity is such that the string exerts no pressure on the tube at one point, it will exert no pressure on the tube at any other point.

The tube may therefore be supposed to disappear then without in any way altering the motion of the string as it passes through the deformation. The deformation in fact becomes a wave which stands still in a moving string. As the motions of the string and the deformation are relative to one another the equation we have found will be true if the string be supposed to stand still and the deformation to move. *A transverse deformation therefore travels along a perfectly uniform and flexible string without change of shape and with a uniform velocity.* Thus if a string has a tension T and its mass per unit length is ρ a deformation will run along it as a wave with a velocity $\sqrt{\dfrac{T}{\rho}}$.

We may use this result to calculate the frequency of the vibration of a string sounding its fundamental tone when its length, its tension, and its mass per unit length are known.

Suppose our deformation to take the shape of an endless sine curve, and suppose its wave-length to be λ and τ to be the time of a complete vibration of this wave-length. It follows that

$$\tau = \frac{\lambda}{v} = \lambda \sqrt{\frac{\rho}{T}}$$

The frequency is the inverse of the time of a complete vibration, for a string vibrating 256 times in a second takes $\dfrac{1}{256}$ seconds to make a vibration. Thus, if n be the frequency of the vibration

$$n = \frac{1}{\lambda} \sqrt{\frac{T}{\rho}}$$

Now suppose this wave motion to meet one of identical shape travelling in the opposite direction. They will have the same velocity and they will combine to produce a stationary transverse wave whose motion would be represented, on a very exaggerated scale, by Fig. 28. The wave-length and frequency of vibration of the stationary wave will be the same as that of the component progressive waves as we can see from Fig. 28. We may regard any two successive nodes as being the fixed ends of a string vibrating so as to sound its fundamental tone. The distance between two successive nodes is half a wave-length. Hence, if l is the length of a string between such fixed ends,

$$\lambda = 2l$$

Therefore, for a string of length l, whose mass per unit length is ρ and which has a tension T, the frequency, n, of a complete vibration of the string in its fundamental mode of vibration is

$$\frac{1}{2l} \sqrt{\frac{T}{\rho}}$$

The frequency varies inversely as the length, inversely as the square root of the mass per unit length of the string, and directly as the square root of the tension. These conclusions can be demonstrated experimentally with a monochord in the manner explained in textbooks on sound. The tension has to be increased fourfold to raise the pitch an octave. To lower it an octave the mass may be increased fourfold. This explains why wire is wrapped round the G string of a violin and the lower strings of a pianoforte. The use of wire in this way tends to avoid the loss of flexibility which would occur if a thick string were used. If a thick string were used it would give unsatisfactory results, for the reason explained in the second paragraph on p. 126, Chapter X.

APPENDIX XI

(See page 88)

A NOTE ON AUDITORIUMS

STUDENTS of music are sometimes expected to be acquainted with the general principles of the acoustical effects which distinguish good auditoriums from bad ones. The general principles themselves, to which this note is confined, are fairly simple. Their application is a matter calling for expert judgement; while the measurement of intensities of sound in auditoriums calls for special knowledge and delicate apparatus. The only satisfactory way of determining beforehand whether a room of a particular design will make a good music-room or concert-hall, or can be modified to improve its acoustic qualities, is to send the architect's designs to a physical laboratory specially equipped to undertake acoustical investigations. The technique employed in such investigation is described in standard books such as *Acoustics of Buildings* by Davis and Kaye.

Every one is familiar with the alteration made to a room when, at spring-cleaning time, its curtains, upholstered furniture, and carpets are removed. The piano gains greatly in power and sonority and the room is no longer dead to sing in as are many drawing-rooms. If anything there is too much reverberation in the now nearly empty room. This indicates the essential problem of the auditorium. Questions of bad echoes may arise, but the most frequent failing of a music-room or large hall is excessive reverberation. Some reverberation is necessary if the room is not to feel oppressively dead to the singer or performer. But excessive reverberation conveys a confusion of sounds to the listener.

Reverberation is due to the continuance of diffused echoes in the room. Sound is a form of energy: it cannot be destroyed. Some of it may escape from the room, through open windows or ventilators. Otherwise the sound can only disappear by being turned into other forms of energy, such as heat, usually through absorption by porous surfaces. In absorbent surfaces the sound vibrations are rapidly converted into heat by friction inside the pores of the surface. For this reason the reverberation of a room is greatly reduced when the carpets, upholstered furniture, and curtains are restored after spring-cleaning is over; and in concert-halls the reverberation is less when the audience is present because all their clothes are absorbent to sound. The remedy for excessive reverberation is to introduce absorbent surfaces. Special plasters which absorb sound are available and can be used to fill panels in the walls and ceiling.

For quantitative data about reverberation we are indebted to the pioneer work of the late Professor Sabine. As a standard, he used sounds about one million times the intensity of sound on the threshold of audibility. This is the intensity of the sound of the voice of a speaker in a large room at a point close to him. Sabine showed that reverberation was practically independent of the positions in the room of the source of the sound and the observer, and that the effect of a given amount of absorbent material was as a rule practically independent of its position. The amount of reverberation which was desirable for music was determined by experiments with a number of persons of musical taste. It may be taken as lying between 1 and 2 seconds; that is, on ceasing at its source, a sound of standard intensity should die away to that of the threshold of audibility in this time. For orchestral music the period of reverberation may well be about 2 seconds. In very large halls a longer period may be preferred, probably because of the improved intensity of the sound. For speech less reverberation is desirable than for music.

Undesirable echoes must be avoided. The shape of the room must not give echoes which concentrate the sound unduly. It is bad to have a dome or a barrel-shaped ceiling for this reason. If a curved ceiling is used its radius should be at least twice the height of the room. A source of sound on the floor-level would then be at the focus of the circular section of the ceiling and sound waves from it would be reflected with a plane wave front. Compare this, for example, with a concave mirror reflecting a light placed at its focus as in Fig. 52.

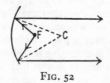

FIG. 52

A coffered ceiling, of a design such as that shown in section in Fig. 53, breaks up the reflections in a satisfactory manner, and if necessary the panels can be filled with plasters or other materials which absorb sound. Walls may be treated similarly to break up unduly long surfaces.

The probable performance of a room in causing echoes can be calculated geometrically from the architect's designs; or model sections can be tested in the laboratory. One method of testing is by the use of the ripple tank. The model section is placed in the surface of water in the ripple tank, and cinema photographs are taken of the reflection of the ripples caused in the surface of the water by a point which is made to dip into the surface of the water and out again. Another method is that of spark photography. Every one is familiar with the quivering effect which air rising over hot ground produces on the appearance of objects seen through it. This is a refractive effect due to the alteration of the density of the air by expansion caused by the heat. It is possible to photograph instantaneously the refractive effect on a ray of light of the condensation caused in the air by a pulse of sound. The sound is made by an electrical discharge from a Leyden jar, and the pulse is photographed by the flash from the discharge of other Leyden jars which is made to happen a moment later by a suitable delay device. Both methods have been employed at the National Physical Laboratory. The technique is fully described in *Acoustics of Buildings* already referred to.

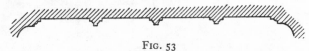

FIG. 53

It is possible to use resonance and forced vibration to strengthen the source of sound. This is one use of a platform. The source of sound and the platform act as a system coupled mechanically when a singer or a piano stands on the platform.

The desirable qualities of an auditorium are summarized as follows by Davis and Kaye in their book on the acoustics of buildings:

(a) there should be no perceptible echoes.

(b) the loudness should be adequate and as uniform as possible throughout the room.

(c) there should be no undue reverberation; each sound should decay quickly enough to be inoffensive when the next sound arrives.

(d) the reverberation should be as uniform as possible for sounds of different frequencies; for any selective absorption of sounds of different frequencies would affect the musical quality of the notes heard by disturbing the balance of the series of partial tones, and a similar result would follow if notes of particular frequencies were intensified by marked and selective resonances.

(e) the walls should be reasonably sound-proof so as to exclude external sounds.

INDEX

The addition of fn. *to a page number indicates that the reference is to a footnote.*